EVERYDAY ENGINEERING

Putting the **E** in STEM Teaching and Learning

Richard H. Moyer and Susan A. Everett

Photography by Robert L. Simpson III

NSTApress

National Science Teachers Association

Arlington, Virginia

National Science Teachers Association

Claire Reinburg, Director
Jennifer Horak, Managing Editor
Andrew Cooke, Senior Editor
Wendy Rubin, Associate Editor
Agnes Bannigan, Associate Editor
Amy America, Book Acquisitions Coordinator

ART AND DESIGN
Will Thomas Jr., Director
Lucio Bracamontes, Graphic Designer, cover and interior design

PRINTING AND PRODUCTION
Catherine Lorrain, Director

NATIONAL SCIENCE TEACHERS ASSOCIATION
Francis Q. Eberle, PhD, Executive Director
David Beacom, Publisher

1840 Wilson Blvd., Arlington, VA 22201
www.nsta.org/store
For customer service inquiries, please call 800-277-5300.

FSC
www.fsc.org
MIX
Paper from
responsible sources
FSC® C005010

NSTA is committed to publishing material that promotes the best in inquiry-based science education. However, conditions of actual use may vary, and the safety procedures and practices described in this book are intended to serve only as a guide. Additional precautionary measures may be required. NSTA and the authors do not warrant or represent that the procedures and practices in this book meet any safety code or standard of federal, state, or local regulations. NSTA and the authors disclaim any liability for personal injury or damage to property arising out of or relating to the use of this book, including any of the recommendations, instructions, or materials contained therein.

eISBN 978-1-936959-79-2

LIBRARY OF CONGRESS CATALOGING-IN-PUBLICATION DATA
Moyer, Richard.
  Everyday engineering : putting the E in STEM teaching and learning / Richard H. Moyer and Susan A. Everett ; photography by Robert L. Simpson III.
    p. cm.
  Articles previously published in Science scope.
  Includes bibliographical references and index.
  ISBN 978-1-936137-19-0
  1. Inventions--Popular works. 2. Engineering--Popular works. I. Everett, Susan A. II. Title.
  T47.M725 2012
  620--dc23
                    2011050981

CONTENTS

PREFACE

THE IDEA FOR the "Everyday Engineering" column in *Science Scope* began with our interest in design and production issues related to the simple ballpoint pen. We were struck by the elegance of the means for retracting the reservoir and transferring the ink to paper. There are many engineering and science concepts involved in these processes. Almost all ballpoint pens work the same way. And some are sold for as little as 19 cents. This led to the development of an activity we used with our students and then a workshop at the Detroit NSTA Regional Conference in the fall of 2007. The pen material was published as "Everyday Engineering: What Makes a Bic Click?" (see Chapter 2) in the April/May 2009 issue of *Science Scope*. Another workshop was presented at a McGraw-Hill Science and Technology Symposium for science supervisors from across the country. The feedback from both teachers and administrators was overwhelmingly positive. We realized that others, too, shared our zeal for appreciating the engineering of simple, common, everyday devices.

For some months, we found ourselves taking a number of things apart—becoming more and more intrigued with the design of the seemingly simple. When thought about in this light, paper clips and pump soap dispensers become fascinating. Learning the history of how these everyday objects were developed is also fascinating. We then proposed to Inez Liftig, the editor of *Science Scope*, a regular feature called "Everyday Engineering" that would investigate the science and engineering of simple, everyday items. This proposal was quickly endorsed by Inez and the *Science Scope*

Advisory Board. In order to provide teachers, scout leaders, workshop leaders, parents, and engineers leading outreach activities with the collection of "Everyday Engineering" ideas, this book was conceived.

We envision that *Everyday Engineering: Putting the E in STEM Teaching and Learning* can be used in a number of different ways. Since the new *A Framework for K–12 Science Education* (NRC 2011) includes engineering as a disciplinary core idea, teachers may use our book to integrate the engineering concepts within their normal science curriculum. Other teachers may wish to teach one or more separate engineering units. After-school group leaders, summer enrichment program administrators, and even youth group leaders may choose to use the activities as well. We also hope that parents who wish to share the world of engineering with their children will find this book to be beneficial since each activity is complete, includes safe procedures, and explains the science and engineering concepts involved as well as the history and development of the everyday object's design.

Richard H. Moyer and Susan A. Everett, October 2011, Dearborn, Michigan.

Reference

National Research Council (NRC). 2012. *A framework for K–12 science education: Practices, crosscutting concepts, and core ideas.* Washington, DC: National Academies Press.

ACKNOWLEDGMENTS

THE AUTHORS ARE indebted to our students for their enthusiastic willingness to engage in the classroom activities we present. The feedback we receive is of great value. We are also appreciative of sixth-grade science teacher Cindy Pentland and her middle-level students in metropolitan Detroit. They were ready to try out our engineering activities on a regular basis and were always eager to get their hands on everyday devices and their minds involved in how things work. In addition, Samantha Hartshaw and Leah Walkuski, two of our student assistants, provided much help working through many of the activities in Ms. Pentland's classroom.

We also want to recognize our colleagues at the Inquiry Institute at the University of Michigan–Dearborn. Specifically, we want to thank Chris Burke, John Devlin, Charlotte Otto, Steve Rea, and Paul Zitzewitz for their willingness to talk with us about our ideas.

Finally, we are indebted to our editors at *Science Scope*: Editor Inez Liftig, Managing Editor Ken Roberts, and Consulting Editor Janna Palliser. Their good work makes us better. They work tirelessly to make *Science Scope* such an outstanding classroom aid to its thousands of readers. We are also grateful for the work done by Claire Reinburg, director of the NSTA Press and her outstanding staff, especially Agnes Bannigan, who was responsible for putting this volume together so seamlessly.

PART 1

What Is This Thing Called Engineering?

CHAPTER 1

AN INTRODUCTION TO EVERYDAY ENGINEERING

WHEN WE WROTE our first "Everyday Engineering" article ("What Makes a Bic Click?"), we began by asking what people usually think of when they hear the word *engineering*. Does it bring to mind bridges, highways, satellites, cell phones, or HD television? According to the International Technology and Engineering Educators Association (ITEEA), "Everyone recognizes that such things as computers, aircraft, and genetically engineered plants are examples of technology, but for most people, the understanding of technology goes no deeper" (ITEA 2007, p. 22). While we certainly agree with this position, our point of view is that engineering is also about the very simple. Every day we use common, ordinary objects such as scissors, can openers, and zippers that involve quite sophisticated engineering.

All of these items that we normally take for granted have, since their invention, gone through numerous design iterations in order to meet some human need or function. As the noted professor of engineering at Duke University Henry Petroski observes in the preface to his well-known book, *The Evolution of Useful Things*, "Other than the sky and some trees, everything I can see from where I now sit, is artificial" (Petroski 1992, p. *ix*). Virtually all of these artificial objects in our environment have been designed by someone in order to solve a problem or serve a human need or want.

Science, Engineering, and Technology

The public's understanding of science, engineering, and technology, and their relationships to one another, are often misunderstood. Science is usually defined as both knowledge of how the natural world works as well as the practices we employ to determine those understandings. The National Research Council (NRC) in the new *A Framework for K–12 Science Education* (2012) defines engineering as "a systematic and often iterative approach to designing objects, processes, and systems to meet human needs and wants" (NRC p. 202). Technology is the product of engineering; "technologies result when engineers apply their understanding of the natural world and of human behavior to design ways to satisfy human needs and wants" (NRC p. 12).

While these three constructs are unique ideas, there are indeed areas of interconnectedness. According to *A Framework for K–12 Science Education*, "It is impossible to do engineering today without applying science in the process, and, in many areas of science, designing and building new experiments require scientists to engage in some engineering practices" (NRC 2012, p. 32). For example, in the early 1600s, Galileo improved the recently invented telescope and with this technology, observed what he first thought were three small stars near Jupiter. In subsequent observations he noted that they were orbiting Jupiter and were indeed moons. Here we see an example of how the science of light and lenses led to technological

development (the telescope), which in turn led to new science (the discovery of Jupiter's moons). Another example of this cross-linking of science and technology is the hundreds of everyday spin-offs from NASA's space exploration. The infrared ear thermometer is one such example. It is based on astronomical technology originally developed for measuring the temperature of distant objects (NASA 2011).

Science and Engineering Education

The interconnectedness of science and technology has resulted in a greater emphasis on the integration of engineering into the teaching of science. To that end, *A Framework for K–12 Science Education* states, "We are convinced that engagement in the practices of engineering design is as much a part of learning science as engagement in the practices of science" (NRC 2012, p. 12). The *Framework* includes engineering in two powerful ways: The knowledge of engineering is one of the disciplinary core ideas along with physical, life, and Earth and space sciences. Again, given their interconnectedness, the practices of engineering are included in the *Framework* parallel to the practices of doing science. The engineering practices begin, for example, with the definition of a problem while the practices of science begin with a question. Similarly, one engineering practice is designing solutions, and the parallel science practice is constructing explanations. Moreover, some practices, such as model building, investigating, and analyzing data, are included in both the practices of science and engineering.

The inclusion of engineering into K–12 science is not a new idea. In the years following Sputnik's launch in 1957, there was grave concern that the United States had fallen behind the Russians in the race to explore and exploit space. The result was a greater emphasis on science and mathematics with the aim of producing more American engineers and scientists. Then, in the 1980s, the American Association for the Advancement of Science (AAAS) began Project 2061,

an effort that was concerned about the scientific literacy of *all* Americans, not just the relatively small number who may pursue careers in science and engineering. They issued the first of several reports, *Science for All Americans* in 1989 and *Benchmarks for Science Literacy* in 1993. Project 2061 developed recommendations for student understandings in science, mathematics, and technology. The following quote, nearly 20 years old, comes from Project 2061—the year of the next expected appearance of Halley's Comet—and recognizes that the effort to change science education is likely to be an ongoing endeavor:

> The terms and circumstances of human existence can be expected to change radically during the next human life span. Science, mathematics, and technology will be at the center of that change— causing it, shaping it, responding to it. Therefore, they will be essential to the education of today's children for tomorrow's world. (AAAS 1993, p. xi)

After a lengthy process, the National Committee on Science Education Standards and Assessment, which was established by the National Research Council, completed work on the *National Science Education Standards* in 1996. This document also includes technology standards alongside the science content. The technology standards consist of two major ideas— (1) "abilities of technological design" and (2) "understandings of science and technology" (NRC 1996, p. 107). The dual foci encourage students to engage in the engineering design process in much the same way that they are encouraged to actively participate in science investigations.

As described above, *A Framework for K–12 Science Education* extends the inclusion of engineering within science education beyond that of earlier science education reform documents. Thus it can be seen that over the last 50 years, the emphasis has evolved from focusing on teaching science for the primary purpose of developing future scientists and engineers to developing a scientifically and technologically literate citizenry. This parallels the explosion of technology in our society.

Informal Engineering Education

Since the inclusion of engineering is relatively new and limited in K–12 education, many have learned about engineering and the design process outside of the classroom. Children may have had engineering experiences through extracurricular activities, such as scouting groups or science clubs. The parents of inquisitive children have long found old clocks or other objects to take apart and rebuild together. Children also may have read some of the books about invention and design such as the very popular *The Way Things Work* (Macaulay 1988). Cable television also has a number of programs on inventions and how things are made. The activities presented in this book may be used in these types of non-school settings as well.

Parents and teachers may also be interested in learning more about the engineering process through additional reading. The periodical *Invention and Technology* specializes in the history of engineering achievements. A popular author of numerous books for the general public on engineering and design is Henry Petroski, a civil engineering professor at Duke University. He has devoted much of his career to helping people appreciate and understand engineering, particularly the design process. His books include such titles as *To Engineer Is Human* (1985) and *The Evolution of Useful Things* (1992).

The Plan of the Book

Each of the lessons in this book was originally published in NSTA's middle-level journal *Science Scope* as part of our bimonthly column, "Everyday Engineering." Our guiding principle for choosing topics was that they had to be "everyday" and have the potential for creating an unique investigation. Our goal was to go beyond activities such as bridge building and egg drops that are available in many publications. We wanted to provide fresh engineering activities to the many teachers who integrate engineering into middle-level science instruction.

In this book, the lessons are organized by topic into five chapters rather than by date of publication. For example, there are four lessons in the Kitchen Engineering section, ranging from pop-up turkey timers to zipper-type baggies. The other sections are Office Engineering, Bathroom Engineering, Electrical Engineering, and Outdoor Recreational Engineering.

Each lesson is driven by an inquiry investigation that integrates core science and engineering concepts. All investigations include some history of the everyday product or design, a student activity sheet, and complete teacher background information. In every case, we have selected materials with cost in mind. Most materials will typically be found in middle-level science classrooms, the school's office supply cabinet, the supermarket, or the dollar store. Finally, each piece includes detailed photography—often a close-up view—to show the operation or characteristics of the particular device. For example, the "Holiday Blinkers" lesson contains photos taken through a digital microscope showing the movement of the bimetallic switch inside the tiny bulb that regulates the blinking function. "An Absorbing Look at Terry Cloth Towels" lesson photos show magnification of loop and cut piles.

Each inquiry investigation follows the learning cycle format, which has a long history of use in science classrooms, dating back to the early 1960s science reform after the launch of *Sputnik* (Atkin and Karplus 1962). The learning cycle has gone through many iterations but probably the most popular is the BSCS 5E Instructional Model (BSCS 1992). The critical point of the learning cycle pedagogy is that students are actively involved in investigations that lead to conceptual understandings—the reverse of traditional classrooms where the conceptual development, usually lecture or reading, comes before laboratory activity of any kind. The 5E Model employed in these lessons include the following stages:

- Engage: The purpose of the Engage stage is to set up an explorable question for the students to investigate.

- Explore: In this stage, students conduct the investigation and collect evidence to support their conclusions.
- Explain: In this stage, students develop conceptual understandings based on the evidence they collected in the Explore stage.
- Extend: Students relate new learning to prior knowledge and real-world applications.
- Evaluate: Assessment can take place at any point of the learning cycle; a summative assessment can serve to guide further teaching and learning.

Readers who wish to have more information on the learning cycle may refer to some of our other publications (see Everett and Moyer 2007 or Moyer, Hackett, and Everett 2007).

Finally, we hope your experiences with this book will not only help with ideas for integrating engineering curriculum into your classroom but also provide the impetus for you, the next time you open a kitchen drawer, to stop and contemplate why the garlic press is made the way it is, who first thought of the idea, and how you might improve on the design. To that end we hope to encourage a sense of engineering literacy—that is, engineering for all.

References

American Association for the Advancement of Science (AAAS). 1989. *Science for all Americans.* New York: Oxford University Press.

American Association for the Advancement of Science (AAAS). 1993. *Benchmarks for science literacy.* New York: Oxford University Press.

Atkin, J. M., and R. Karplus. 1962. Discovery or invention. *The Science Teacher* 29 (2): 121–143.

Biological Sciences Curriculum Study (BSCS). 1992. *Science for life and living.* Dubuque, IA: Kendall Hunt.

Everett, S., and R. Moyer. 2007. Inquirize your teaching. *Science and Children* 44 (7): 54–57.

International Technology Education Association (ITEA). 2007. *Standards for technological literacy: Content for the study of technology.* 3rd ed. Reston, VA: ITEEA.

Macaulay, D. 1988. *The way things work.* Boston: Houghton Mifflin.

Moyer, R., J. Hackett, and S. Everett. 2007. *Teaching science as investigations: Modeling inquiry through learning cycle lessons.* Upper Saddle River, NJ: Pearson/Merrill/Prentice Hall.

National Aeronautics and Space Administration (NASA). 2011. Infrared ear thermometers. NASA @ Home and city. NASA. *http://www.nasa.gov/externalflash/nasacity/index2.htm*

National Research Council (NRC). 1996. *National science education standards.* Washington, DC: National Academies Press.

National Research Council (NRC). 2012. *A framework for K–12 science education: Practices, crosscutting concepts, and core ideas.* Washington, DC: National Academies Press.

Petroski, H. 1985. *To engineer is human: The role of failure in successful design.* New York: St. Martin's Press.

Petroski, H. 1992. *The evolution of useful things.* New York: Vintage Books.

PART 2

Office Engineering

CHAPTER ②

WHAT MAKES A BIC CLICK?

WHAT DO YOU think of when you hear the word *engineering*? Does it bring to mind bridges, highways, satellites, cell phones, or HDTV? According to the International Technology and Engineering Educators Association (ITEEA), "Everyone recognizes that such things as computers, aircraft, and genetically engineered plants are examples of technology, but for most people, the understanding of technology goes no deeper" (ITEA 2002, p. 22). While we certainly agree with this position, our point of view is that engineering is also about the very simple. Every day we use common, ordinary objects, such as scissors, can openers, and zippers, that involve quite sophisticated engineering. The following lesson describes how basic engineering ideas can be integrated with other subjects, in this case some basic concepts of the properties of matter.

The ballpoint pen is another ideal example of simple engineering that we use every day. But is it really so simple? The ballpoint pen is a remarkable combination of technology and science. Its operation uses several scientific principles related to chemistry and physics, such as properties of liquids and simple machines. While there are differences in pen construction and price, most pens actually work the same way. They represent significant advancements in the engineering development of writing instruments—and most can be purchased for less than a dollar.

Historical Information

Around 700 AD, the quill pen was first used and remained the main writing instrument for over a thousand years until the invention of the more practical fountain pen in the late 1800s (Bellis 2006; Peeler 1996). Fountain pens frequently leaked, and the ink smeared easily before it dried on paper. In the 1930s, Hungarian László Bíró thought the quick-drying ink used to print newspapers might work in fountain pens. It proved to be too thick. Therefore, he designed a rolling ball at the end of a tube—an idea actually first patented 50 years earlier. Ballpoint pens became popular during the 1940s—pilots used them in WWII. The pens could write for up to a year without having to be refilled. However, their overall quality was poor and after an initial fad, fountain pens continued to dominate the market. Around 1950 in France, Marcel Bich introduced the inexpensive yet reliable Bic ballpoint that became the most popular pen in Europe and later worldwide. Ballpoint pen technology evolved to include a retractable tip, various-size roller balls, gel-type inks, colors, and ergonomic barrel designs. As a teacher, you may choose to discuss this information with students as needed, or it can be easily located by having students conduct an internet search using the search terms *ballpoint pens* or *history of writing implements*.

FIGURE 2.1 Parts of a capped pen

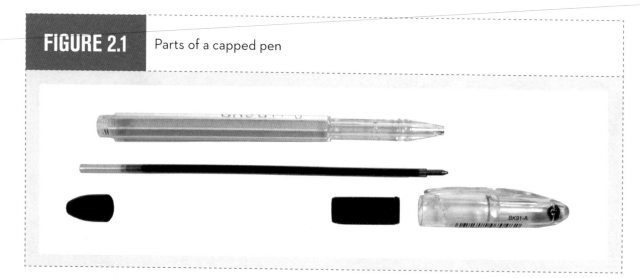

Investigating Ballpoint Pens (Teacher Background Information)

Engage

Safety note: Students must wear goggles.

Gather an assortment of ballpoint pens. Have a short discussion with students to give them opportunities to share some of their prior knowledge and experiences to lead to the explore question for the lesson: How do the parts of the two types of pens (capped and retractable) compare?

Explore

Distribute a baggie of several capped and retractable "clicker" pens to each group of students with enough pens for each pair of students to work with one pen of each type. A class of 30 would require 15 of each type of pen. You can often find inexpensive or used pens at thrift stores or dollar stores. As the pens will not be destroyed, they can be used again by other classes of students. Simply have students put the pens back together when the Explore phase is completed. You may wish to have some extra pens in case some parts are lost or damaged in the process. Have students work in pairs to carefully take apart the capped pen. Students should examine the capped pen and describe the system that keeps the cap attached. Repeat the above steps with the retractable pen. Figures 2.1 and 2.2 show the disassembled pens (we selected pens that are clear so the parts are easily seen).

Explain

Students should organize their findings in a table and answer the questions. You should decide if your students will need a blank table or will be able to design one for themselves. Figure 2.3 (p. 12) shows our results for this activity and is provided here as an aid to the teacher. Results may vary slightly, depending on the actual pens used. Then, students share answers to begin a discussion of the activity. Using a diagram may be helpful to explain the function of the various parts. The following information on pens will help teachers lead the discussion: Essentially, all pens with a cap have some feature whose purpose it is to keep the cap in place. In the simplest case, part of the barrel near the tip is made slightly larger than the internal diameter of the cap. The cap deflects as it is pushed onto the barrel. Another design employs a cone shape near the end of the barrel and a ridge inside the cap. Finally, some pens use a ring around the barrel and a corresponding ridge inside the cap that stretches over the ring on the barrel and snaps into place.

Note in Figure 2.2 the two-piece push-button assembly. The top piece protrudes out of the end of the

FIGURE 2.2 Parts of a retractable pen

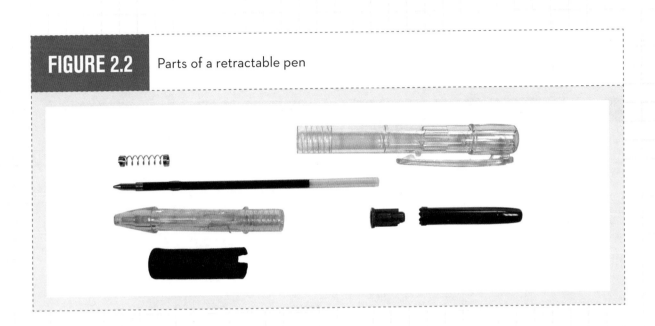

barrel. It has ridges at the bottom that move through guide rails inside the upper portion of the barrel. Each time the push button is depressed, the bottom piece is rotated because of the slant at the end of the guide rails. Thus, every time the button is depressed, the pen's tip protrudes or retracts. A rather sophisticated system built for less than a dollar! Students should determine that the spring is what holds the tip of the pen inside the barrel. If they have difficulty with this idea, have them think about whether the spring is being compressed or stretched (it is in compression), and how it therefore exerts a force on a small flange on the reservoir, pushing it toward the top of the barrel.

Extend

Students examine a different retractable pen that does not use a push-button clicker. One screw-type pen per group should be sufficient and can be reused many times. Some pens, for example, use a screw-type mechanism to raise and lower the tip of the pen. Often, these are slightly more expensive than the push button type. A spring is not necessary for this mechanism.

Evaluate

Students might show their understanding of this technology by designing an innovative way to improve

on the basic design of a pen. They should draw their design and explain how each part works as part of the overall system. Students may suggest designing pens that have additional features, such as pens that write in different colors, lights for writing in the dark, erasable pens, pens for writing on different surfaces, a music-playing pen, and so forth. Most of these pens may already be available in the marketplace, but you may have a student who designs a pen that is truly innovative (see Activity Worksheet 2.1, p. 14).

Investigating Ink With a Model Bottle Pen (Teacher Background Information)

Engage

Safety note: Students should wear goggles during this investigation and be cautioned not to taste the "inks." Because food coloring can stain clothing, students should wear an apron and cover the work area with newspaper.

Discuss with students how the ink in ballpoint pens does not leak out of the reservoir when it is not in use, but still manages to be able to write a smooth line as soon as it is moved across a piece of paper (see Activity Worksheet 2.2, p. 15). The ink in ballpoint pens must

FIGURE 2.3   Pen parts comparison

Part	Retractable	Capped
Cap		X
Barrel	X	X
Pocket clip	X	X
Spring	X	
Push-button assembly	X	
Ink reservoir	X	X
Ink-reservoir collar	Never	Sometimes
Roller-ball tip	X	X
End cap		Usually
Cushioned grip	Sometimes	Sometimes

FIGURE 2.4   Ink results

Water	Alcohol
Skips	Skips
Blobs	Fewer blobs
Slow drying time	Very fast drying time
Leaks a little	Leaks significantly more

have properties that allow it to flow easily from the reservoir, but also be of sufficient consistency to not clog or leak. Further, the ink in ballpoint pens dries almost instantly when applied to the paper. However, it must not dry while in the reservoir. Students will investigate the following question: How does the type of ink affect its performance?

Explore

In order to investigate how a ballpoint pen transfers ink to paper, we made a model out of a small bottle with a roller-ball applicator. Such bottles are easily obtainable online and are often used for application of perfumes and the like. They can be purchased for less than 15 cents each. Once the bottles are filled with "ink" they can be used over and over again by multiple classes. Have students work in pairs with two bottles. Afterward, the bottles can be stored and reused. We have not found an everyday substitute, as roll-on deodorant bottles are expensive and removing the roller ball is quite difficult. Students will use the roller-ball bottles to simulate a pen and test different inks. (A basic investigation might compare water and

a few drops of food coloring with rubbing alcohol and food coloring. You may wish to have students select additional properties of ink, such as thickness, to explore. To explore thickness, prepare several bottles with corn syrup diluted with water in different proportions, such as 10%, 50%, and 90%.) To assess performance of the different inks, students may want to consider drying time, smearing, skipping, leaking from the reservoir, or overall quality of the line produced. Depending on your students, you may wish to provide the format of the data table or have students design the data table.

Explain

Our ink results are shown in Figure 2.4. In a sense, students are comparing a major difference between older water-based, fountain-pen ink and more modern ballpoint pens that use either oil- or alcohol-based solvent. The alcohol "ink" is less cohesive than the water and thus leaks more readily (Figure 2.5). For the same reason, it produces fewer "blobs," because it is more adhesive to the paper and therefore flows more evenly. The ball is the key to transferring the ink from the reservoir to the paper. Close inspection of the ball reveals small indentations that help collect ink as the ball turns and transfers it to the paper. The ink that is stored in the reservoir must be thick enough so that it does not flow out around the ball when the pen is not in use.

Extend

The lesson can be extended by having students explore other variables, including the type of paper, the application of pressure, the type of line—straight or curved—and the speed with which they attempt to write. Or you may choose to have students test other types of ink as suggested in Explore.

Evaluate

At the end of the lesson, students should be able to determine some of the primary issues a pen designer needs to consider when selecting ballpoint pen ink. A possible answer is as follows: The ink must flow easily from the reservoir but also be of sufficient consistency to not clog or leak. It also must dry almost instantly when applied to the paper but not dry out while in the reservoir.

Conclusion

Even though students are familiar with and make use of cell phones, MP3 players, laptop computers, video games, and so on, they also still make use of everyday, low-tech devices such as ballpoint pens. Few students likely understand, or have even thought about, the engineering that is involved in designing a pen. Working with such low-tech devices affords the opportunity to develop an understanding of some core concepts of engineering.

FIGURE 2.5 Student comparing model

References

Bellis, M. 2006. A brief history of writing instruments. *http://inventors.about.com/library/weekly/aa100197.htm.*

International Technology Education Association (ITEA). 2002. *Standards for technological literacy: Content for the study of technology.* Reston, VA: ITEA.

Moyer, R., J. Hackett, and S. Everett. 2007. *Teaching science as investigations: Modeling inquiry through learning cycle lessons.* Upper Saddle River, NJ: Pearson/Merrill/Prentice Hall.

Peeler, T. 1996. The ball-point's bad beginnings. *Invention and Technology* 11 (3): 64.

ACTIVITY WORKSHEET 2.1   Investigating Ballpoint Pens

Engage

Look at the assortment of ballpoint pens and think about the characteristics of each pen. In this activity, you will investigate the following: How do the parts of the two types of pens (stick and retractable) compare?

Explore

Safety note: Wear chemical-splash safety goggles.

1.   Take apart the capped pen without breaking the pieces. Do not try to take the ink tube apart. Make a drawing of how the parts fit together. What is the purpose of each part?

2.   Now, take apart the retractable pen. Make a drawing of how the parts fit together. What is the purpose of each part?

Explain

1.   Put your findings into a table to compare the parts of the two types of pens to answer the question from Engage: How do the parts of the two types of pens (stick and retractable) compare?

2.   Look at your drawing of the retractable pen. Does the spring keep the tip inside the pen or does it push it out when you want to write?

Extend

1.   Look at another retractable pen to examine the parts.

2.   How does this retractable pen compare to the first one? Explain how this retractable pen works without a "clicker."

Evaluate

Now that you have examined many different pens, consider a way that you could improve or invent a new pen. Draw your design and explain how each part of the pen works.

ACTIVITY WORKSHEET 2.2 — Investigating Ink With a Model Bottle Pen

Engage

Think about writing with a ballpoint pen. The ink flows out when writing but does not leak out when it is not in use. What properties do you think the ink needs? How fast does it need to dry? In this activity, you will investigate the following question: How does the type of ink affect its performance?

Explore

Safety note: Wear chemical-splash safety goggles.

1. Brainstorm with the class on how to test the performance of a pen.
2. Using two different types of liquids in the model bottle pens, test each pen by writing on plain white paper. Observe the qualities of the writing from each pen. Record your observations in a data table.

Explain

1. Using your observations, answer the question from Engage: How does the type of ink affect its performance?
2. Look at the roller ball on the model pen with a hand lens. What do you notice? How do you think the "ink" is transferred from the bottle to the paper?

Extend

1. Brainstorm with the class some other factors that you could test with the model bottle pens.
2. Select one factor to investigate. Record your observations. Share your results with your classmates.

Evaluate

Now that you have investigated with the ink, consider some of the issues that a pen designer needs to think about when selecting ballpoint pen ink. What properties does the ink need to have?

CHAPTER 3

CLIPS AND CLAMPS

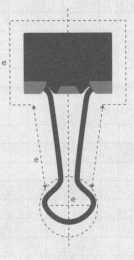

IF YOU OPENED a desk or junk drawer, how many different kinds of clips or clamps would you find? You are likely to find paper clips and binder clips of several different sizes each, and perhaps a chip clip, clothespin, or hair clip or two. Around your house, you may also find a clipboard, alligator clips on a wire, or jumper cables and spring clamps in your garage. If you look at this collection of clips and clamps, you will notice that they all function by means of a spring. Springs are not always coils of wire but can be any device that has the tendency to return to its original position when displaced. Other than the paper clip, most of the clips or clamps you have around school or your home will probably fall into two different types of designs. One uses a coiled spring (Figure 3.1, p. 18) that holds the jaws together, and the other uses either a triangular or C-shaped piece of sheet metal or plastic as a spring (Figure 3.2, p. 18).

In this 5E Model lesson, students will investigate the design of simple, everyday binder clips and how they function. In the process, students will apply concepts related to levers and forces as noted in the middle-level National Science Education Standards (NRC 1996, p. 154). The lesson also addresses the following ITEEA standard for middle-level students: "A product, system, or environment developed for one setting may be applied to another setting" (ITEA 2002, p. 49). To this end, students will examine a number of clips in order to determine how they function and will experiment with altering one of the parameters (i.e., the length of the handle).

Historical Information

People have been holding papers together since at least the 13th century. Early methods included sewing papers together with string and ribbon. When straight pins were invented, they were employed to hold paper together. Then, in the mid to late 1800s, various types of paper clips became available. One advantage of paper clips was that they could hold papers together without making holes.

The binder clip was introduced around 1911 by a teenager named Louis E. Baltzley (Hales 2006). He was motivated to invent the binder clip because his father was a writer and had many papers that needed to be held together. The binder clip's basic design is still in use today, 100 years later.

Paper fasteners are so common that it is difficult to completely pin down their development over the years. As noted by Henry Petroski, an engineering professor at Duke University: "When all is said and done, any attempt to sort out the origins and patent history of the paper clip may be an exercise in futility. For there appear to have been countless variations on the device, a great multiplicity of forms, and some of the earliest and most interesting versions seem not to have been patented at all" (Petroski 1994, p. 62).

It is often compelling to be able to share exact dates and inventors with students, but technology does not always develop singularly but rather on many fronts. In this case, there was a need for holding papers together, and when the machinery was developed to bend wire economically, an explo-

FIGURE 3.1   Examples of spring-type clamps

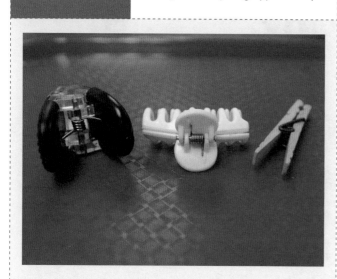

FIGURE 3.2   Examples of binder clamps

sion of ideas ensued. You may wish to share early paper clip images from the Early Office Museum website with your students (*www.officemuseum.com/paper_clips.htm*).

Investigating Clips and Clamps (Teacher Background Information)

Materials

Safety note: You must provide students with safety goggles for this entire investigation in the unlikely event that a clip breaks loose and becomes a projectile.

For this activity, you will need an assortment of binder clips as well as various other types of clips, such as hair clips, food bag clips, clothespins, spring clamps for carpentry, or a clipboard. Containers of assorted metal binder clips can be either found in your school's office supply area or purchased for approximately $6 for a box of 60. You will also need three to four clips of the same size for each group. We recommend using the 1.9 cm (¾ in.) size, which are available for approximately

$4.50 for a box of 36. The clips can be reused from one class to the next.

For the Engage stage, each group of three to four students will need two identical binder clips, one of which has the handles removed. For the Explore stage, each group will need three additional 1.9 cm clips, which have had their handles changed. It is recommended that the teacher prepare these clips prior to class. Use the assorted clips to obtain handles of different sizes and transfer them to the 1.9 cm clips. We used one smaller handle, one normal handle, and two larger handles (Figure 3.3). The exact lengths are not important and will likely vary from one manufacturer to another. Students will also need two unsharpened pencils, duct tape, scissors, and a ruler. For the demonstration in the Explain stage, the teacher will need two clips from the Explore stage—the standard size and the largest size—duct tape, and three or four identical textbooks. Each group will need several different types of clips in the Extend stage. Finally, for the Evaluate stage, students can manipulate an array of different-size binder clips (with standard handles attached), see

FIGURE 3.3 Four 1.9 cm clips with different-size handles

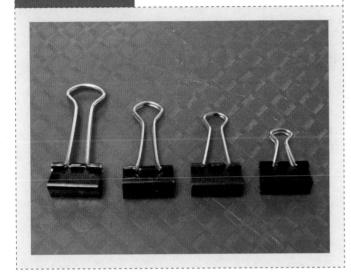

FIGURE 3.4 Clip with extra-long handles

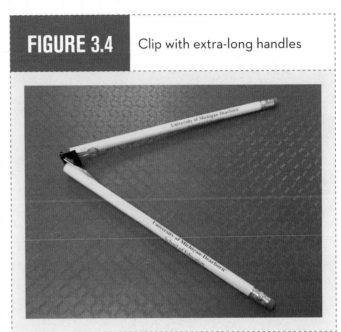

a photo, or observe the image of clips projected from an overhead.

Engage

To activate prior knowledge, initiate a brainstorming session with students regarding their experiences with clips and clamps. Focus attention on the binder type of clip and encourage students to determine how it works by opening and closing. Give students a clip with the handles removed and let them feel how difficult it is to open without the handles—in fact, it may not be possible to open. At all times during the investigation, students should not try to force clips apart with tools of any kind. If enough are available, each student should have a clip to examine. Have students share their ideas and lead the discussion to the following Explore question: How does the size of the handle affect the opening and closing of a binder clip?

Explore

In this investigation, the size of the binder clip is held constant at 1.9 cm, while the length of the handle is varied (i.e., the independent variable). The dependent variable, how difficult it is to open each clip, is qualitatively measured by squeezing the clip with the fingers. Students will likely rank the clips or state that some are "easy" and some are "more difficult." Students should conclude that the longer the handle on a clip, the less force is required to open it. While the same applies to closing the clip if it is done slowly—with continuous pressure applied—students will likely simply let the clips snap closed and thus report (correctly) that the length of the handles is not a factor in closing. When students design a clip that requires less force to open, they will likely construct one with much longer handles, perhaps by attaching pencils to the existing handles (Figure 3.4).

Explain

If students are familiar with levers, this Explain stage may be a review and an application. Otherwise, you may need to guide them in understanding how a lever works (e.g., effort and resistance forces) and identifying its main parts (e.g., the fulcrum). In the case of the

FIGURE 3.5 Quantitative testing of handle lengths

FIGURE 3.6 Exploded view of C-shaped spring clips

binder clip, note that there are actually two levers, one on the top and one on the bottom. In the following discussion, we focus on just one of the levers. When opening a clip, the fulcrum is located at the point where the handle touches the top edge of the V-shaped clip. This is the balance point about which the handle rotates as it is squeezed. Thus, the length of the handle (i.e., the length of the lever's effort arm) is the distance the handle extends out past the clip's top. The resistance force is applied by the spring to the end of the handle where it fits into the round grooves. Thus, the resistance arm is from these grooves to the fulcrum at the top end of the clip. The longer the effort arm relative to the resistance arm, the less force will be required to open the clip.

To further this concept, the teacher should conduct the following demonstration. Measure the length of the effort arm on a clip with the standard handle (ours measured 18 mm). Then measure the effort arm on a clip with the largest handle (ours measured 28 mm). Tape the first clip securely to the table and add textbooks until it begins to open (Figure 3.5). While the weight of the books will vary, we selected books

so that it took three books to open the standard size. Repeat using the clip with the longer handle. If you used clips with handles like ours, you should find that it only takes two books to open this clip with the longer handle.

Students should be able to conclude that these quantitative data support what they discovered qualitatively in the Explore stage. Help students understand that because both clips have the same size spring, the resistance (i.e., the force to open the spring) is identical. However, the clip with longer handles requires only two books to exert enough force to open the clip; it takes three books to apply the same force to open the clip with the standard handles. For this reason, we consider the mechanical advantage (MA) of the longer-handled lever to be 3/2, or 1.5. This is equal to the ratio of the effort arms:

$$28 \text{ mm} \div 18 \text{ mm} = 1.56$$

The MA can be determined either way. You may wish to have students calculate the MA of the extended-handle binder clip they designed. The image in Figure

| FIGURE 3.7 | Different-size binder clips |

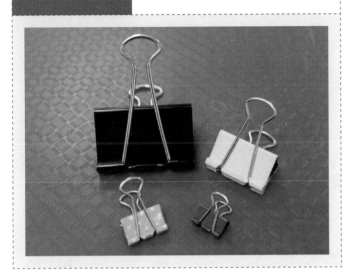

3.4 shows an effort arm of 184 mm compared to the standard arm of 18 mm:

$$MA = 184 \text{ mm} \div 18 \text{ mm} = 10.2$$

Extend

Depending on what is available, students will likely sort their clips into two or three groups: the "binder clip" group, a group using coiled springs, and possibly a third group such as food bag clips. Some food bag clips contain a plastic or metal *C*-shaped spring clip and function the same as the metal binder clips or the *C*-shaped metal binder clips (Figure 3.6). Note that in some binder clips, the spring is plastic rather than sheet metal. If it is part of your curriculum, you may wish to have students identify the fulcrum, effort, and resistance arms of the different types of clips.

Evaluate

Provide students with different sizes of binder clips or show photos (for example, Figure 3.7). Ask why different-size clips have handles of different lengths.

Mechanical Advantage

Consider the force required to open the binder clip. It is nearly impossible to pull a clip apart with your fingers. The use of the handles provides what is called a mechanical advantage, allowing you to open the clip with considerably less force. How is this possible? First, you must consider that you "cannot get something for nothing," as the saying goes. In order to open the binder with the handle, you have to move the handle much farther than the jaws of the clip actually open.

This is because no machine can produce more work than is put into the machine. You may recall that scientists usually define *work* as the product of a force acting over some distance. The smaller force we apply with our fingers acting over a greater distance is equal to the larger force exerted by the jaws multiplied by the smaller distance that they move when they open.

The ideal mechanical advantage of a lever can be calculated by comparing the distances from the fulcrum to the effort force and to the resistance force. If the so-called effort arm is twice as long as the resistance arm, then the ideal mechanical advantage of that lever would be two. In this lesson's Explore stage, making the effort arm (the handle of the binder clip) longer by a ratio of 3:2 resulted in a mechanical advantage of 1.5 for the clip with the longer handles. Because of friction, the actual mechanical advantage will be less than the ideal mechanical advantage and can be determined empirically by calculating the ratio of the actual output force to the input force.

Students' answers should indicate that they understand that longer handles require less effort force to open a given clip and that larger clips with stronger springs require a greater effort to open and thus need longer handles. To test their ideas, students should suggest an experiment similar to the teacher demonstration that measures the effort force needed to open the clips.

Some students may wonder why all clips simply do not have longer handles to make them easier to

open. This question leads to some of the engineering variables that constrain the manufacture of the clips. Some of these factors may include the cost of the additional material, the cost of shipping and packaging larger clips, as well as the bulkiness of the final product. The modern binder clip is another example of everyday engineering in which a design has withstood the test of time. Today's clips are essentially the same as they were 100 years ago. The usefulness of this type of spring has been applied to many different products, such as hair clips, food bag clips, and spring clamps of many types. The simple binder clip is an example of a lowly, yet elegant design used decade after decade. Binder clips are all based on the idea of a simple spring that is used to hold something closed. Perhaps you can think of some types of springs that hold things open.

References

Hales, L. 2006. A big clip job? Think Washington. *Washington Post.* May 20.

International Technology Education Association (ITEA). 2002. *Standards for technological literacy: Content for the study of technology.* 2nd ed. Reston, VA: ITEA.

National Research Council (NRC). 1996. *National science education standards.* Washington, DC: National Academies Press.

Petroski, H. 1994. *The evolution of useful things: How everyday artifacts—from forks and pins to paper clips and zippers—came to be as they are.* New York: Vintage Books.

Resources

Early Office Museum: History of the paper clip. *www.officemuseum.com/paper_clips.htm*

Moyer, R. H., J. K. Hackett, and S. A. Everett. 2007. *Teaching science as investigations: Modeling inquiry through learning cycle lessons.* Upper Saddle River, NJ: Pearson/Merrill/Prentice Hall.

ACTIVITY WORKSHEET 3.1   Investigating Clips and Clamps

Engage

Safety note: You must wear safety goggles at all times during this investigation.

1. Brainstorm a list of different types of clips and clamps with which you are familiar and discuss with the class.

2. Carefully examine the binder clip your teacher has provided to determine how it works. How does it open? What keeps it closed? Record your ideas.

3. Now examine the clip without handles. Compare how easily it opens to the one with handles. Record your findings.

4. In the Explore stage that follows, you will test clips with handles of various sizes. How does the size of the handle affect the opening and closing of the clip?

Explore

1. Your teacher will provide your group with binder clips for your investigation.

2. Measure the length of each handle and then carefully try to open each binder clip using only your fingers. Construct a data table to record your results.

3. Qualitatively describe how much force was required to open each binder clip and record.

4. Using the materials provided by your teacher, construct a binder clip that requires much less force to open. Make a drawing of your idea. Share your plan with the teacher before you build your binder clip.

5. Test your idea and record your results.

6. Use your results to answer the following questions: How does the size of the handle affect the opening of the clip? How does the size of the handle affect the closing of the clip?

Explain

1. Share your results with your classmates.

2. Why do you think the length of the handle is important? How is the handle like a lever?

3. Your teacher is going to do a demonstration with two clips that have handles of different lengths to quantitatively determine how much force is needed to open each clip. Compare the length of each handle to the amount of force needed to open each clip. Determine the length of the handle that protrudes out beyond the clip. Record your results in the table on page 24.

(Activity Worksheet 3.1 continues)

Activity Worksheet 3.1 continued

Force needed to open clips with different handles		
	Length of handle protruding beyond the clip (cm)	Number of books needed to open the clip
Clip 1		
Clip 2		

4. How do the data you collected in the table relate to what you discovered in the Explore stage?
5. Think about how many books are needed to open each clip. How much easier was it to open one of the clips?
6. The mechanical advantage (MA) of a lever is a measure of how much the effect of the applied force is increased. What is the MA of the binder clip with the longer handle?
7. What do you notice about the ratio of the length of the longer handle to the smaller one? Compare this to the MA you determined in Step 6.

Extend

1. Look at the different clips your teacher has provided to determine how they open and close.
2. Sort the clips based on their operating mechanism.
3. Which are similar to binder clips and which are different? Explain.

Evaluate

Different-size clips have different-size handles. Why? How would you set up an experiment to test your idea?

PART 3

Kitchen Engineering

CHAPTER 4

TIME'S UP, TURKEY— POP-UP THERMOMETERS

DRIED OUT, OVERCOOKED turkey is a sure way to ruin a festive dinner. This disappointing bird likely results from an awareness of the dangers of consuming undercooked poultry. The United States Department of Agriculture recommends turkeys reach an internal temperature of 165°F (74°C) in the thickest part of the thigh (USDA 2008). Meat thermometers can be awkward to place due to the abundance of bones. Because of these problems, each year 30 million Thanksgiving turkeys have a built-in thermometer that pops up when the turkey is properly cooked.

Turkey timers are an example of how engineering solved a common, everyday problem. The ITEEA recommends that students in grades 6–8 understand the following technology standard: "New products and systems can be developed to solve problems" (ITEA 2002, p. 27). The 5E Model activity, as with all of the activities in this book, integrates this engineering concept with science content, in this case, to investigate the development of the disposable pop-up cooking thermometer. Possible related science content includes heat transfer and melting points of various materials.

Historical Information

In the 1960s, the California Turkey Producers Advisory Board wanted to respond to complaints of dried-out turkeys and to help people avoid overcooking them at home. One of the Board members, Eugene Beals, said, "Why don't we find some sort of gadget, something to stick in [the turkey] and tell when the turkey is done?" (Taylor 2005, A32) As they sat in a meeting room, someone looked up at the fire sprinklers on the ceiling and wondered if the same system that turned the sprinklers on could be used for a turkey timer. Fire sprinklers at that time were turned on when a low-melting temperature alloy was heated sufficiently by flames in a room, causing a water valve to open. Beals and his colleagues spent nearly one year finding the right material to melt at the appropriate temperature to ensure a tasty turkey. The patent has been sold, but the basic principle is still being used today. In less than 50 years, more than two billion timers have been used in turkeys and turkey products.

All pop-up thermometers (i.e., cooking timers) work essentially the same way. As can be seen in Figure 4.1 on page 28, the pop-up timer is made of an indicator stem in a housing. The tip of the stem is embedded in an alloy typically containing bismuth, lead, and cadmium, which melts at a given temperature—about 180°F (83°C) for turkeys. When the turkey is heated to this temperature, the alloy melts, the stem is released, and a spring pushes it so that it "pops up," and the cook knows the turkey is the proper temperature (and not overcooked). Pop-up timers can now be purchased with different alloys that melt at different temperatures for use in beef roasts, hams, and other meats.

FIGURE 4.1 Detail of pop-up turkey thermometer

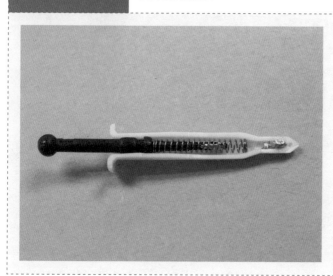

FIGURE 4.2 Heating turkey timer in water

Investigating Pop-Up Thermometers (Teacher Background Information)

Safety note: Students must wear chemical-splash goggles and thermal aprons for this activity.

For each group of two to four students, you will also need a hot plate, an oven mitt, a small metal pan (or beaker) with water, a thermometer, tongs, and a turkey timer (available online; two for $3). You could also ask students and other teachers before Thanksgiving to bring in their used timers after the holiday. In this case, wash, sanitize in boiling water, and reset as discussed below. Students will also need some ice or cold water to cool their water between trials in the Explore phase.

Engage

Initiate a discussion about how students' families cook turkey. How long does it take? How do you know when it's done? If no one mentions a turkey pop-up timer, show one to the class. Ask students to predict at what temperature a turkey timer will pop up. After several predictions, ask students to share ideas for how to set up an investigation to test their prediction.

Explore

Safety note: Because water is involved, be sure to plug the hot plate into a ground fault circuit interrupter (GFCI) outlet.

Students can determine the temperature at which the turkey timer pops up by placing a pop-up timer in water heating on a hot plate. Remind students to avoid touching the timer to the bottom of the pan. While monitoring the increasing temperature of the water, students will record the temperature at which the timer activates (Figure 4.2). This will occur around 181°F (83°C). Have students fill their pans about half full of water. This allows room for them to add ice or cold water between trials in order to cool the water to about 140°F (60°C). After the timer pops, students should leave the timer in the hot water for at least another minute. This is to assure that the alloy holding the stem fully melts. Students need to push the stem back into the housing immediately and hold it securely for two minutes (Figure 4.3). This is to allow the alloy adequate time to cool and resolidify with the stem embedded in it. If students are not successful in getting their stem

FIGURE 4.3	Resetting turkey pop-up thermometer

FIGURE 4.4	Cork model: Unpopped and popped

to remain in the down position, they need to reheat it and try again.

Explain

In this stage of the model, students share their findings. Each group might compare the range and the mean temperature at which the pop-up turkey thermometer activated. Our turkey timers activated at about 181°F (83°C). Ask students to share their thinking on how the timer works. You may wish to share the story of how the idea was originally conceived—when someone looked at a fire sprinkler and realized that the valve was held shut by a piece of metal that melted during a fire. In this way, the sprinkler "sensed" when the temperature in a room was too hot. Explain that the pop-up turkey thermometer senses the temperature at which a turkey is properly cooked in a similar manner. You may want to ask students to explain the transfer of thermal energy. Thermal energy is transferred from the oven to the turkey (and the alloy in the timer), causing temperatures to increase. When the melting point of the alloy is reached, it melts and releases the embedded stem.

When reheated, the timer process can be reversed and the stem can be pushed back into the alloy. Students need to understand that the timers usually activate at the same temperature, which indicates that the alloy melts at the same temperature each time. Not all materials melt at a given temperature; some, like waxes, melt over a range of temperatures.

Extend

In this stage, students design and build a model of a pop-up timer. Figures 4.4 and 4.5 (p. 30) show two different examples (if you do not have access to a refrigerator, you may choose to end this lesson at this point). These are just two possibilities. Encourage your students to come up with other designs by providing an assortment of materials. For example, freezing a toothpick in an ice cube can make a rudimentary pop-up. Figure 4.4 shows a pop-up timer made out of a small plastic vial weighted with washers, so it will sink when placed in a larger beaker of water. The indicator stem is made of a cork and a brass brad. The brad ensures that the cork will float in an upright position. The cork

FIGURE 4.5 — Pen model: Unpopped and popped

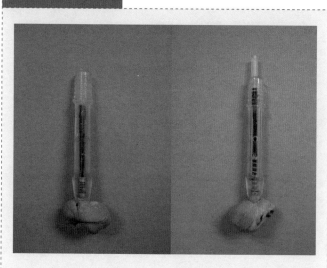

is pushed down into the water-filled vial with the brad nesting in the washers. A piece of cellophane tape holds the cork flush with the top of the vial, and the entire model is placed in a freezer. When the water freezes, the stem is embedded in the ice, and the tape can be removed. To test the model, it can then be placed in a larger beaker of warm tap water (approximately 131°F or 55°C), which will melt the ice and the cork stem will pop up. Using the vial shown in Figure 4.4, we found it would activate in two or three minutes.

The second model, shown in Figure 4.5, is made with the barrel of a retractable ballpoint pen. The ink reservoir must be trimmed so that when pushed into the barrel, compressing the spring, it is flush with the end. The barrel is inserted into a piece of modeling clay, sealing the end and providing ballast, allowing the system to sink in a beaker for testing. Fill the barrel with water and again use cellophane tape to compress the spring and hold the ink reservoir in place during the freezing process. To test the model, remove the tape and place it in a larger beaker of warm tap water. We found this model to activate quite quickly—in less than 20 seconds.

Evaluate

Students should compare their model timer to those of other groups. Assign each student a model from another group to analyze. Each student should make a sketch and label the key component parts of the real turkey timer—some sort of housing, a stem, and the melting material in which the stem is embedded. This analysis and sketch allows you to assess whether students are able to apply their new knowledge of the timers. Students' explanations should describe that once the ice melts, it releases the stem, which is then free to either float up or be pushed up (above the housing) with a spring.

Conclusion

Engineering often involves a great deal of creativity in order to develop new products to solve everyday problems. The pop-up turkey thermometer is an ideal example of this creative process: Technology that was developed for fire control was cleverly applied to the very different problem of overcooked turkeys. Helping students understand this aspect of engineering may encourage them to consider a career in an engineering field.

References

International Technology Education Association (ITEA). 2002. *Standards for technological literacy: Content for the study of technology.* 2nd ed. Reston, VA: ITEA.

Moyer, R., J. Hackett, and S. Everett. 2007. *Teaching science as investigations: Modeling inquiry through learning cycle lessons.* Upper Saddle River, NJ: Pearson/Merrill/Prentice Hall.

Taylor, M. 2005. *Eugene Beals—Inventor of the turkey pop-up timer.* San Francisco Chronicle. October 9.

United States Department of Agriculture (USDA). 2008. *Food safety and inspection service: Appliances and thermometers. www.fsis.usda.gov/Fact_Sheets/Kitchen_Thermometers/index.asp.*

ACTIVITY WORKSHEET 4.1   Investigating Turkey Pop-Up Thermometers

Engage

1. Predict the temperature that the pop-up turkey thermometer will activate.
2. In the next part of this activity, you will investigate the following: Do pop-up turkey thermometers consistently activate at the same temperature?

Explore

Safety note: You must wear chemical-splash goggles and a thermal apron. One student in each group will need an oven mitt.

1. Place a pan less than half full of water on a hot plate. One member of your group should hold the thermometer in the water so that it does not touch the pan. Constantly observe the change in temperature.
2. Using tongs, another group member should hold the pop-up turkey thermometer so that the tip is in the water as it is heated.
3. Record the water temperature when the stem pops up.
4. Hold the timer in the hot water for another minute. Then remove it from the water and (with the oven mitt) push the stem back into the housing and hold it without moving for two minutes. Turn the heat down on the water at this time.
5. After two minutes, the stem should remain in the housing.
6. Repeat steps 1–5 two more times. Construct a data chart and record the temperature at which the timer activates for each trial.

Explain

1. Find the mean activation temperature for your group. What is the range of your data?
2. Share your findings with other groups.
3. How do your data compare with other groups? What is the mean activation temperature for the entire class? What can you infer from this piece of information?
4. Answer the question from Engage: Do pop-up turkey thermometers consistently activate at the same temperature?
5. How do you think the pop-up turkey thermometer works?

Extend

1. Using the available materials, your group will design and build a model of a pop-up timer.
2. Draw a sketch of your plan. Your plan must include the same components as the actual pop-up turkey thermometer: a housing, a stem, and a material that melts. In this model, you will use ice instead of an alloy. Your model may or may not make use of a spring.
3. Write an explanation and draw a sketch of how your design works.
4. Show the sketch to your teacher for approval.
5. Build your model and test it in a beaker of warm tap water. (Safety note: You must wear chemical-splash goggles during this activity.)

Evaluate

1. Examine other groups' model timers.
2. Make a sketch and write an explanation of how one of them works.

CHAPTER 5

CHARCOAL—CAN IT CORRAL CHLORINE?

HOW MANY WAYS have you used water today? Have you ever wondered who used that water before you—and for what purpose? Because our water is used over and over, a faucet has been described as the other end of someone else's drain. Water is obviously one of our most precious resources, and we have a limited supply available for our use. For this reason, the United Nations proclaimed 2011 as the International Year of Chemistry and, as part of that designation, invited teachers worldwide to participate with their students in the Global Water Experiment (IYC 2010). Students had the opportunity to take part in four different water-related investigations to assess water quality and then share data with other students around the world. The project began on World Water Day (March 22) and ran throughout 2011. This chapter looks at a related water exploration—designing charcoal filtration systems like those used in filtering water pitchers, some coffee pots, and refrigerators. There are several curriculum tie-ins for this lesson; most notably is natural water filtration using different types of sediment, such as sand or gravel.

In this 5E Model activity, students will investigate how charcoal (carbon) filters can be used to remove chlorine from water. They will design a filtering system to reduce the chlorine concentration in a sample of water while also minimizing the amount of charcoal that is required. The ITEEA's middle-level standard for attributes of design states, "Requirements for design are made up of criteria and constraints" (ITEA 2002,

p. 95). In this lesson, the design requirement is to reduce the presence of chlorine in a water solution. The constraints include the time required for filtration and the amount of charcoal needed, which is a function of the system's total cost.

Historical Information

Early Sanskrit writings from the 15th century BC indicate that people attempted to treat and filter their drinking water. The walls of ancient tombs from this time show drawings of the purification of drinking water (Figure 5.1, p. 34) (Baker 1981). Chlorination of water began in the 1800s and is still widely used to purify water today, as it destroys many harmful organisms. Charcoal filters also have a long history: "The efficacy of powdered charcoal in preventing or removing bad tastes and odors from water, and therefore clarifying it, was established experimentally by Johann Tobias Lowitz in 1789–90" (Baker 1981, p. 26). By the end of the 19th century, sand and crushed coal were used to filter drinking water in the United States (Jesperson 1996a). Modern carbon filtration systems first came on the scene in the late 1960s, resulting in the plethora of water filters we have today, including water pitchers, faucet filters, and even water-bottle filters. In 1966, Heinz Hankammer started producing Brita charcoal filters in Germany, naming his company after his young daughter (Clorox Company 2011).

FIGURE 5.1

Early water purification device (Baker 1981)

Investigating Charcoal Filters (Teacher Background Information)

Materials

For the Engage stage, try to locate a few common household water filters you may have on hand such as those used in filtering water pitchers, refrigerators, water faucets, and coffee pots, or you can search for online images by using a phrase such as *carbon filters*. For each group of three to four students, you will need a funnel, several pieces of filter paper, and three small beakers. If funnels and filter paper are not available, you can substitute a half-liter empty water bottle (a 1 L bottle cut in half) with the top inverted and inserted into the bottom to form a funnel and receptacle with a coffee filter inside. In addition, each group will need a 300 ml solution of bleach and water. The dilution students are working with is very weak, but you should review the material safety data sheet (MSDS) for bleach (Fisher Scientific 2008).

If you have six groups per class and five classes, you will need approximately 9 L of bleach solution. To make the solution, add 9 ml of (5.25%) bleach to 9 L of water. This will result in a solution that will test at approximately 6 mg/L of chlorine. Each group will also need 45 cm^3 of rinsed activated charcoal. The activated charcoal is available at stores that sell pet fish and aquarium supplies. It can be purchased for about $8 per kg, which should be sufficient for five classes of six groups. You can clean all of the charcoal at one time by placing it in a plastic tub and then rinsing until the water flows clean and the charcoal particulates have been removed. Hold your hand over the top of the tub to prevent the charcoal from being washed away as the water overflows the tub. Chlorine test strips are available for approximately $17 for 100 strips from restaurant and pool supply companies as well as pet stores selling aquarium supplies. For students to design their filtration systems, have available supplies such as coffee filters; funnels; a mortar and pestle; clear, colorless, plastic tubing (about 1 cm in diameter; the type used for aquariums works well); and support devices such as ring stands and clamps.

For the Extend stage, you will need a collection of household filters (or photos) such as a filtering water pitcher (about $12 for a 5-cup pitcher [1250 ml]), refrigerator cold-water filter, water-faucet filter, and so on. In addition, you might want to have an impurity on hand of such concentration that it cannot be removed by quickly passing through a filter. A 15% solution of vanilla in water is one such example. About 400 ml of the vanilla solution will be needed for a teacher demonstration. If each group will do this, you will need about 100 ml per group, or 2.4 L for five classes of six groups.

It is useful to have an aquarium filter to display or a photo of one for the Evaluate stage.

Engage

Safety note: While the concentration of bleach in the solution used by students is very low, students should wear indirectly vented chemical-splash goggles, aprons, and gloves (vinyl or nitrile) while conducting this investigation. Although there may be adequate ventilation in the science laboratory, it is recommended that this activity be done under a fume hood. If some students are exhibiting breathing difficulties, this is a symptom of inadequate ventilation. Some students may be sensitive to chloramines. Check with the school nurse in advance for students with allergies or asthma.

Review salient safety points noted on the dilute chlorine bleach MSDS with students prior to working with the hazardous chemical. Remind students that diluted bleach on the skin can cause severe irritation and if splashed in the eye can cause blindness. Make sure there is a 10-second access to both an eyewash station and shower in case of a splash incident. Also, remind students that bleach will discolor clothing. Upon completing the activity, students should wash their hands with soap and water.

Discuss with students household filters with which they are familiar, such as coffee pots/filters (French press coffee pots), sieves, colanders, and tea infusers for brewing loose-leaf teas. They might mention automobile oil and gasoline filters, furnace air filters, or dust masks. You might also want to discuss the history of water filtering and purification at this point. Students will likely note that some household water filters make use of charcoal. Ask students if they know what charcoal is. They may not be aware that it is essentially the element carbon.

Ask students to share their ideas of the purpose of coffee filters. What do they allow to pass through and what do they filter out? Students should predict what happens when a bleach solution is poured through a paper filter and should give reasons for their predictions. They can test the filtrate by using a chlorine test strip. Follow the directions on the package, which instructs students to dip the strip into the water (and perhaps wait up to a minute) and then compare the color of the strip to a standard that is on the package to determine the concentration of chlorine remaining in the solution. Many students may be surprised to learn that the solution has barely changed after passing through the filter.

FIGURE 5.2 Filtering bleach solution and test strips

Most public water supplies in the United States have chlorine added, primarily as a disinfectant to kill many harmful organisms. It is also helpful for the removal of certain odors and unpleasant tastes (Jesperson 1996b). The amount of chlorine present in most drinking water ranges from 0.2 to 1.0 mg/L. Your local water could contain more chlorine than this, however, and still be potable—the World Health Organization guideline is at least 0.5 mg/L and less than 5 mg/L (WHO 2003). More chlorine is added at the treatment facility because a portion combines with nitrates and organic materials and is therefore unavailable for disinfecting. Some individuals find the smell and taste of chlorinated water objectionable. However, students should be reminded that it is perfectly safe to drink chlorinated water from a municipal water system. Some people choose to filter the chlorine from the water in their homes either at the tap or by using a filtering water pitcher. If your local water is chlorinated, students can see if they can detect its odor. You may want to have students test your local water with a test strip, but bear in mind that the chlorine concentration may be less than can be measured.

FIGURE 5.3 One possible filtering design

Explore

Students will begin the exploration by pouring 100 ml of the dilute bleach solution through 45 cm³ of charcoal pellets in a filter (Figure 5.2) and then test the filtrate for chlorine with a test strip. In this case, students will discover that little of the chlorine or odor is removed. Challenge students to hypothesize why one pass through a carbon filter was not totally successful in removing the chlorine and brainstorm possible solutions. Students may conclude that to remove impurities, one must increase the contact between the solution and the charcoal. This can be done either by increasing the amount of time the solution is in contact with the charcoal or by increasing the amount of charcoal through which the solution must pass. This choice essentially frames the engineering challenge of this exploration. If students have difficulty coming up with possible solutions to this challenge, reinforce that the two primary variables that can be readily manipulated are (1) the time the solution is in contact with the charcoal and (2) the amount of charcoal used in the filter. Students will

design charcoal filters that remove the most chlorine and odor, bearing in mind the engineering constraints of time to filter and the amount of charcoal required. One technique they might try is to pour the solution through the same charcoal pellets several additional times. Another technique is to grind the pellets in a mortar and pestle (students will already be wearing goggles) in order to increase the surface area of the charcoal in contact with the bleach solution. Students could also pour the solution over some pellets in a beaker, let it sit for a few minutes, and then decant the liquid. Passing water through a plastic tube filled with charcoal also increases the exposure of the solution to the carbon (Figure 5.3).

Explain

Have students share their engineering solutions. All of the methods described above can be used to remove most of the chlorine from the water, and students may even obtain a 0.0 mg/L reading as measured with the test strip. Discuss students' ideas for determining which design would be the most efficient. The amount of charcoal required will depend on the design selected. For example, less charcoal might be needed if one pours the solution through the filter more times. However, that would increase the time needed. Or, less charcoal is also needed if it is ground. But then again, it will take a bit longer for the solution to filter through.

There are two reasons carbon is an effective filtering medium: (1) it is quite porous and thus presents a large surface area to trap particulates, and (2) it has an affinity for many molecules that, essentially, are more attracted to it than to water. That is the case with chlorine in water, and many other contaminants, as well. If you have chosen to do this activity in conjunction with a larger unit on natural water filtration, you will want to make those connections at this point.

Extend

Pour 100 ml of the dilute bleach solution through a filtering water pitcher (Figure 5.4) and then use a

FIGURE 5.4
Filtering water pitcher and water bottle

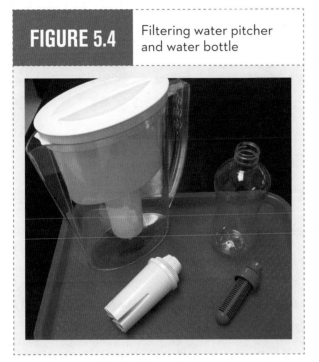

FIGURE 5.5
Filter containing granulated charcoal

test strip to measure the chlorine concentration in the filtrate. You will find that most commercial pitchers will reduce the chlorine significantly—perhaps even to zero as measured by the test strips. Most commercial filtering pitchers have a reservoir on top with a carbon filter through which tap water passes into a lower reservoir. Water is then poured out of the lower part of the pitcher. The filter contains granulated charcoal, thus increasing the surface area over which the water passes (Figure 5.5). Note also that the filter is cylindrical in shape, which increases the contact between the water and the charcoal; both of these design elements reduce the amount of charcoal required. Students should also note that most other charcoal water filters have a similar shape for these same reasons.

You may wish to have students try to filter other contaminants. If you leave a vanilla solution in a beaker of charcoal overnight, both the odor and the color will be removed. You might also try a solution of vegetable dye.

Evaluate

If you have an aquarium in your classroom that uses a charcoal filter, have students observe its operation. If not, obtain a picture of one online. Students should conclude that the water is moving through the charcoal filter rapidly and continually, thus increasing the contact of the water with the charcoal. Perhaps this is similar to one of the designs created by one of the student groups (i.e., pouring the solution through the pellets several times).

Conclusion

Most engineering designs seek an optimal solution to several constraining variables. In this case, the variables were the time it takes to filter a quantity of water as well as the cost involved with the amount of charcoal required. When engineers design a water pitcher, they must also take into account an appropriate size for pouring and storing. They must consider the ease of changing the filter and whether the pitcher can be

manufactured and sold at a profit. After completing this activity, students should appreciate that there is never just one possible solution to any design problem. Rather, there are likely to be several options available to minimize constraints and maximize performance.

References

Baker, M.N. 1981. *The quest for pure water*. Vol. 1, 2nd ed. Denver, CO: American Water Works Association.

Clorox Company. 2011. About Brita, our history. *www. brita.com/about-brita*.

Fisher Scientific. 2008. Material safety data sheet: Bleach. *http://fscimage.fishersci.com/msds/91020.htm*.

International Technology Education Association (ITEA). 2002. *Standards for technological literacy: Content for the study of technology*. 2nd ed. Reston, VA: ITEA.

International Year of Chemistry (IYC). 2010. The global water experiment. *http://water.chemistry2011.org/web/iyc/home*.

Jesperson, K. 1996a. Search for clean water continues. *www.nesc.wvu.edu/old_website/ndwc/ndwc_DWH_1. html*.

Jesperson, K. 1996b. Safe water should always be on tap. *www.nesc.wvu.edu/old_website/ndwc/HistSafeWater. html*.

World Health Organization (WHO). 2003. Chlorine in drinking water. *www.who.int/water_sanitation_ health/dwq/chlorine.pdf*.

Resources

Dacko, M. 2004. Tried and true: Inquiring about water quality. *Science Scope* 27 (9): 34–36.

Moyer, R., J. Hackett, and S. Everett. 2007. *Teaching science as investigations: Modeling inquiry through learning cycle lessons*. Upper Saddle River, NJ: Pearson/Merrill/Prentice Hall.

Orna, M.V. 1994. History: On the human side in solubility and precipitation. Stillwater, OK: Oklahoma State University. *http://intro.chem.okstate.edu/ChemSource/Solubility/page17.htm*

Young, R., J. Virmani, and K. Kusek. 2001. Creative writing and the water cycle. *Science Scope* 25 (1): 30–35.

ACTIVITY WORKSHEET 5.1 Investigating Charcoal Filters

Engage

Safety note: Throughout this entire investigation, be sure to follow appropriate safety guidelines regarding observing liquids. Do not taste any liquids in the lab. You should wear chemical-splash safety goggles, aprons, and gloves while working with the chlorine solution and work only in a well-vented room.

1. Brainstorm with your classmates and make a list of different types of filters that people use. What are the purposes of the different filters? How are they all the same? What are some differences that you see?

2. Do you have any water filters at home? If so, describe what kind you have. How do you think they affect your drinking water?

3. Chlorine is added to most municipal drinking water. Why do you think this is done? Your teacher will provide you with 300 ml of water that has been mixed with bleach to represent chlorinated drinking water. Test the bleach solution with a chlorine test strip and record the amount of chlorine indicated by the color code.

4. What do you predict will happen to the bleach solution after it passes through a paper filter? The liquid that passes through a filter is called the filtrate, and anything that does not is called a residue. Pour the bleach solution through the paper filter, and then use a test strip to determine the chlorine concentration. How do your results compare to your prediction?

5. Often, charcoal is used as a filtering medium. In this exploration, you will design a charcoal filter to remove chlorine from water.

Explore

1. Add about 45 cm³ of charcoal pellets to your filter and pour through 100 ml of the bleach solution. Test the filtrate with another test strip. Record the amount of chlorine remaining in the solution.

2. Your engineering challenge is to design a filter that removes the most chlorine from your sample. The constrain-

ing variables to consider in your design are the amount of time it takes to filter the water and the amount of charcoal needed. Using the materials provided, develop a plan with your group and then make a sketch to show to your teacher for approval.

3. Construct your charcoal filter and test its effectiveness by filtering 100 ml of bleach solution. Record your data and draw conclusions about your design. Record the amount of charcoal used, the time required, and the level of chlorine in the filtrate.

Explain

1. Share the results of your group's filter with your classmates, noting similarities and differences. What factors do you think affect the chlorine removal of the filtered bleach solution? Explain.

2. How can you make comparisons with other groups' designs that take into account the cost effectiveness of the different filters? Hint: How much charcoal was used?

Extend

1. Observe the charcoal filter in a filtering water pitcher. Your teacher will pour 100 ml of bleach solution through the pitcher and use a test strip to measure the amount of chlorine remaining in the filtrate. How effective is the filtering water pitcher compared to the filter you designed?

2. Look at ads online for refrigerator, faucet, or whole-house water filters. What do you notice about the shapes of these filters? Based on your exploration, what is your explanation for the shape of the filters in these designs?

Evaluate

In an aquarium filter, the water flows through quickly. Why do you think this is so? Based on what you have learned so far, what is the key to the design of the aquarium filtering system? Write an explanation of how you think aquarium filters work. You can include a drawing.

CHAPTER **6**

WHAT MAKES A BETTER BOX?

EVERY MORNING, MANY Americans start their day with a bowl of cereal. Some spend time while they eat breakfast reading the back of the cereal box, but few consider its size, shape, and construction, or realize that it was designed by an engineer. Packaging engineering deals with all aspects of how a product safely and efficiently gets from where it is produced to the end consumer. Many factors are involved: the cost of manufacturing the package; the amount of raw materials needed; the amount of waste created by the production process; the disposal, reuse, or recyclability of the materials after use; and finally, the marketing appeal of the package itself. Often some of these issues are at odds with one another and must be reconciled. The middle-level standards of the ITEEA note, "In order to recognize the core concepts of technology, students in grades 6–8 should learn that trade-off is a decision process recognizing the need for careful compromises among competing factors" (ITEA 2002, p. 38–39). This ITEEA standard will be the focus of this lesson.

Packaging engineers must address the cost of the packaging, as well as the product's image conveyed by the graphics and text on the packaging. As a result, virtually all packages are mini-advertisements for the product they contain. In addition, as a result of consumer advocacy, packaging engineers also consider the effects of their packages on the environment and are attempting to incorporate greener practices.

This is a lesson in which students design, build, and critique cereal boxes. The lesson follows the 5E Model and incorporates engineering and measurement, with a focus on mathematics and the relationship between surface area and volume.

The relationship of surface area to volume is an important idea in many fields of science (e.g., retaining and dissipating heat, mass compared to cross-sectional area in plants and animals, diffusion in cells, and respiration and circulation in more complex organisms). Heat transfer is an important topic at the middle level (NRC 1996, p. 155) and lends itself to surface-to-volume considerations. A classic example is how a dog or cat curls up into a ball when sleeping. This shape minimizes the animal's exposed surface area, thus reducing its surface-area-to-volume ratio, and, therefore, the rate at which it transfers heat to the surroundings. However, if the ratio of surface area to volume is larger, such as in an infant, then the cooling (or heating) is much greater. A small child loses thermal energy at a much greater rate than an adult. An opposite case would be elephants' ears, which serve in part as cooling organs. Their large surface area maximizes the transfer of heat to the elephants' surroundings. In physical science,

consider the fins on an automobile radiator; they increase the surface area to maximize the cooling of the radiator. For more information on teaching about issues of scale, see Everett and Otto (2009).

Historical Information

Cereal, as a breakfast food, has been around since the 1860s but became popularized when W.K. Kellogg accidentally developed flaked cereal in the 1890s. While preparing breakfast food for hospital patients in Battle Creek, Michigan, Kellogg left a pot of wheat cooking overnight. The next day he rolled out the cooked wheat and found it formed flakes. Later Kellogg discovered that corn produced tastier flakes, and a new kind of cereal was born—corn flakes. Former patients asked to buy the flakes by mail. Because of this demand, in 1906 the W.K. Kellogg Company was formed and corn flakes were made commercially available (Kellogg Co.).

The method for packing cereal has not changed much since the early 1900s. The cardboard box has been in use since then, with the inner, waxed liner appearing shortly thereafter. While the vast majority of cereal boxes are still much like the ones of a hundred years ago, there are some new developments. Kellogg is now test marketing corn flakes in smaller, squat-tier, and deeper boxes that take up less shelf space in grocery stores and during shipping (Packaging Digest 2009). Archer Farms is packaging cereals in a recently developed cereal box with rounded side edges, which eliminates the need for the plastic bag liner and closes with a resealable plastic lid (Hofbauer 2008).

It is possible to recycle many types of plastic bags, including the bags inside cereal boxes. The State of New York has recently enacted a law that requires retailers, such as grocery stores, who distribute plastic bags to also collect them for recycling. Most of the recycled plastic bags are used to make imitation wood for construction and for garbage bags (New York State Department of Environmental Conservation). Even though there have not been many changes in cereal packages, both parts (the box and the liner) can be recycled.

Investigating Cereal Boxes: What Makes a Better Box? (Teacher Background Information)

Engage

Have students bring to class a variety of empty cereal boxes. You will need about three different-size boxes

FIGURE 6.1 Sample cereal box data

Cereal	Height	Width	Depth	Surface area*	Volume*
Cereal X	28.6 cm (11.25 in.)	19.7 cm (7.75 in.)	5.1 cm (2.0 in.)	1,619.5 cm² (250.4 in.²)	2,873 cm³ (174 in.³)

* Due to rounding, some measurements may be approximate.

FIGURE 6.2	Equations for volume of largest cube

Volume

height × width × depth

Surface area

2 (height × width) + 2 (height × depth) + 2 (width × depth)

Calculating the volume of the largest cube

$V_{CubicBox}$
17.8 cm × 17.8 cm × 17.8 cm = 5640 cm³ (7 in. × 7 in. × 7 in. = 343 in.³)

The surface area of this cube is

$SA_{CubicBox}$
6 × 17.8 cm × 17.8 cm = 1901 cm² (6 × 7 in. × 7 in. = 294 in.²)

The surface area of the scrap is

SA_{Scrap}
2.5 cm × 35.6 cm = 89 cm² (1 in. × 14 in. = 14 in.²)

And if the 1.25 cm (half-inch) strips are added to the sides, then the volume is

$V_{tallerbox}$
17.8 cm × 17.8 cm × 19.05 cm = 6036 cm³ (7 in. × 7 in. × 7.5 in. = 367.5 in.³)

has the largest volume. Keep in mind that perception can be misleading, as the tallest boxes may not always have the largest volume. A slight increase in the depth of a box can have a surprising effect on volume. Two boxes we measured were the same in all dimensions except one was 0.64 cm (¼ in.) deeper. Its volume, however, was 328 cm³ (20 in.³) larger.

Have students fill each box with a lightweight material, such as foam packaging peanuts or popped corn, in order to visualize the different volumes of the boxes. (Remind students they are not to eat the popcorn!) Students may note that the weights listed on the packages differ widely. This is because some cereals are much denser than others. For example, granola cereals are denser than puffed rice cereals. Cereals are sold by weight rather than volume.

Next have students measure and calculate the volume and surface areas of the three boxes (see Figure 6.1 for sample data). Depending on the math backgrounds of your students, you may need to give them the formulas for each (Figure 6.2).

Students will find that while the sizes of the boxes vary, the basic shape of most cereal boxes is the same—a thin, rectangular solid. Have students share any prior knowledge they have as to why this may be so. The analysis of the relationship of the surface area and the volume is complex if both are allowed to vary, as is the case with cereal boxes. Therefore, in our activity, we hold the surface area constant and have students study the relationship between box shape and volume.

Explore

Safety note: Students should wear safety glasses or goggles when working with scissors.

For each group of three to four students, you will need cellophane tape, scissors, ruler, markers, and a 55.9 × 35.6 cm (22 × 14 in.) piece of poster board, which is exactly one half of a standard-sized sheet. Challenge students to design a box that will have the greatest volume. Encourage students to be creative—a key characteristic of engineering that is not always

for each group of three to four students. The boxes may be reused from one class to another. The volumes of our boxes ranged from approximately 1,802 cm³ (110 in.³) to 5,735 cm³ (350 in.³). The surface areas ranged from approximately 1,161 cm² (180 in.²) to 2,129 cm² (330 in.²). Students need to predict which of the boxes

FIGURE 6.3   Planning the cereal package

FIGURE 6.4   Constructing the box

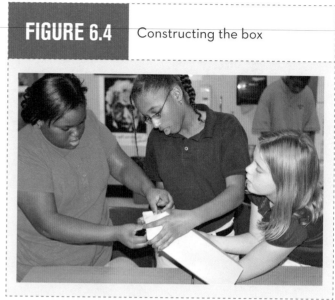

recognized by nonengineers. Give students ample planning time to think about various possibilities before constructing their boxes (Figures 6.3 and 6.4). Some students may have the misconception that the shape of the box has no bearing on its volume. Therefore, not all will recognize that, given a constant surface area, a cube has the largest possible volume for a rectangular solid. If all shapes are considered for a given surface area, a sphere has the greatest volume.

The largest cube that can be made from a 55.9 cm × 35.6 cm (22 in. × 14 in.) sheet has sides of 17.8 cm (7 in.). The six sides can be laid out as shown in Figure 6.5. This results in a 2.5 cm × 35.6 cm (1 in. 1×4 in.) piece of scrap. Some students may cut this scrap into four 1.25 cm × 17.8 cm (0.5 in. × 7 in.) pieces to make a slightly larger box that is 1.25 cm (0.5 in.) taller than a cube. Figure 6.2 shows how the volume of the largest cube is calculated (note that there are some rounding errors when comparing metric to customary measurements).

This box has the largest volume that can be made with the given surface area. Again, a more concrete way to help students visualize the volume would be to fill their boxes with foam packaging peanuts or popped corn using a measuring cup or graduated cylinder.

Explain

Have students share their boxes in order to understand how different shapes affect volume (Figure 6.6). Students should conclude that the cube-shaped boxes maximize the volume for any given surface area. Boxes that approach a cubic shape have a greater volume than tall, skinny boxes, which have the least volume. Students should also notice that none of the commercial boxes are cubic in shape but are, for the most part, tall and skinny. Some students may try other shapes such as the envelope package shown in Figure 6.7 on page 46.

Extend

Have students brainstorm a list of criteria describing the "best" cereal box. Encourage students to explain their definitions of *best*. Help them realize that what one considers the best box may depend on point of view. In engineering, choices must be made among competing factors: "When trade-offs are made, there is a choice or exchange for one quality or thing in favor of another" (ITEA 2002, p. 39). By having to choose

FIGURE 6.5	Layout of cubic box

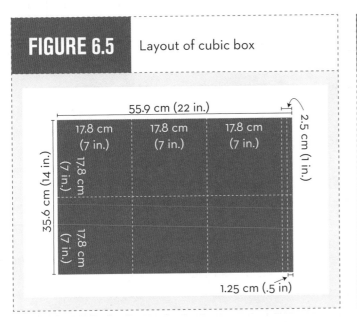

FIGURE 6.6	Comparing boxes

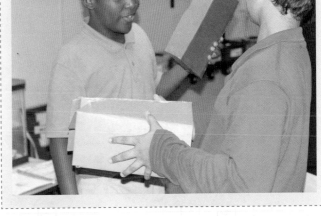

against competing factors in the activity, students will better understand the ITEEA core concept of trade-offs.

For example, from the point of view of the cereal manufacturer, the "best" cereal box might include cost and ease of manufacturing, warehousing and shipping the package, and maintaining the product's freshness. It would be important to be able to manufacture a box with little scrap. While cubic boxes hold the largest volume (for a given amount of cardboard), they may not be the most durable shape for a cereal box, and they are difficult to hold and pour with one hand. The box could be made cheaper without the wax paper liner, but the cereal would quickly become stale. Shipping the cereal from the factory to warehouses and stores is also a concern. Rounded packages would require larger shipping crates and add to expense.

If, however, you are concerned with the marketing and sales of the cereal, you may prefer a tall, skinny box that holds less and requires a greater surface area, using more cardboard and therefore costing more to make. A taller box gives the appearance of holding more than a short, squatty cubic box of the same volume. Thus, by spending a little more on cardboard (rather than on cereal), less cereal appears to be more. The tall, skinny box also provides more space for the product name and promotional information to make a greater impact while displayed on the grocery shelf.

For some people the greatest concern is the amount of resources and energy used to make the package and how it is disposed of or recycled. From the point of view of an environmentally conscientious consumer, the best box is probably quite different—in fact it might be no box at all. Some cereals are now being sold in plain plastic bags. If packaged in a box, perhaps to prevent the cereal from damage during shipment, environmentalists would likely prefer the cube.

Evaluate

After building a container and learning from other groups, students will use this knowledge to create a design for a cereal package. Students will need to wrestle with differing viewpoints, as noted above and in the ITEEA standards presented here, including trade-offs. In their responses, students should discuss the advantages of their design as well as the trade-offs they considered.

FIGURE 6.7 Envelope-shaped cereal package

Conclusion

Most people do not think of package design when they consider the field of engineering. Like all engineering, however, packaging engineering involves design, cost analysis, production, use, and disposal. Creativity and resourcefulness are traits needed by engineers in order to develop alternative solutions to everyday problems. When students have opportunities to experience and develop these traits, they may become more interested in the field of engineering.

References

Everett, S., and C. Otto. 2009. Giants don't exist in the real world: Challenges of teaching scale and structure. *Science Scope* 33 (4): 40–43.

Hofbauer, R. 2008. Package of the year awards: Designs innovate both outside and in. Food and Beverage Packaging. *www.foodandveveragepackaging.com.*

International Technology Education Association (ITEA). 2002. *Standards for technological literacy: Content for the study of technology.* 2nd ed. Reston, VA: ITEA.

Kellogg Co. A historical overview. *www.kellogghistory.com/history.html.*

Moyer, R. H., J. K. Hackett, and S. A. Everett. 2007. *Teaching science as investigations: Modeling inquiry through learning cycle lessons.* Upper Saddle River, NJ: Pearson/Merrill/Prentice Hall.

National Research Council (NRC). 1996. *National science education standards.* Washington, DC: National Academies Press.

New York State Department of Environmental Conservation. Frequently asked questions about plastic bag recycling. *www.dec.ny.gov/chemical/50063.html.*

Packaging Digest. 2009. Kellogg saves space with slim cereal box. *www.packagingdigest.com/article/CA6640968.html.*

ACTIVITY WORKSHEET 6.1   Investigating Cereal Boxes: What Makes a Better Box?

In this activity, you will examine a number of different cereal boxes and then design and build your own.

Engage

1. Examine the cereal boxes provided.
2. Predict which box has the greatest volume.
3. Measure the volume of each box with the materials provided.
4. Use a ruler to measure the height, width, and depth of each box. Record the measurements in a table and then calculate the total surface area of each box. Calculate the volume, as well.
5. Compare the volumes and surface areas of each box. How would you describe the shape of each box? Share your ideas.

Explore

1. Your teacher will give you a piece of poster board. Design a box using this poster board that will hold the greatest amount of cereal. Record how your group decided what shape box to build.
2. Use as much of your poster board as possible to minimize the amount of scrap in your manufacturing process.
3. You can use tape so that you do not need to overlap the edges of the box. The lid should be hinged so that it can be opened to fill the box and measure its volume.
4. Calculate the volume and surface area of your box. How much scrap material did you have?

Explain

1. Share your box and results with the rest of the class.
2. Compare your box to some of the commercial boxes that you observed earlier.
3. What are the properties of the boxes with the greatest volume? Record how the boxes are alike and how they are different.
4. If you built another box, how would you change your plan?

Extend

1. Engineers must consider many factors when they design a package. So far you have only considered two factors—the volume and the surface area. Brainstorm some additional factors to consider when deciding on what makes the overall best cereal box.
2. Consider the pros and cons of how to package cereal and share with the class.

Evaluate

Assume the role of a packaging engineer. Your job is to present a new cereal package to your client, the Better Breakfast Cereal Company. What are the advantages of your new design? What trade-offs are involved with your package? Write a paragraph describing why the company should buy your package.

CHAPTER 7

IT'S (ZIPPED) IN THE BAG

HOW MANY TIMES have you used a plastic baggie in the last week? They are ideal for storing food and marinating before cooking. Because of security regulations, you likely carry your toiletries in a baggie when you fly. Many teachers use baggies to organize hands-on manipulatives; perhaps you keep small parts, such as jigsaw puzzle pieces, in one. We are using them with our students in a garden project to protect tablet computers from wet, soiled hands.

Plastic baggies, especially the sealable variety, seem to be a ubiquitous part of our modern lifestyle. The zipper-type seal on plastic baggies is an example of an unappreciated engineering accomplishment that is relevant to everyday life. In fact, the challenge of fastening two pieces together probably dates back to early humans' making of tools and clothing. The seal on a plastic baggie illustrates the creativity engineers used to find a very simple, yet elegant, solution to a problem. In the lesson that follows, students will investigate different types of sealing mechanisms on plastic bags and then design their own sealing mechanism. Students will also compare a sealable, zipper-type baggie to an actual zipper. While the slider serves a similar function in both, the actual connecting mechanisms are quite different; baggies use a rib and a slot that fit together by friction, and the zipper uses teeth that resemble a hook and socket. The ITEEA's standards for middle-level students state that "technology is closely linked to creativity, which has resulted in innovations" (ITEA 2002, p. 28).

Historical Information

The first plastics date to the 1860s, when cellulose was invented by Alexander Parkes. One of the first uses of plastic was to replace ivory in the manufacture of billiard balls (American Chemistry Council). Plastic wrap for food storage was introduced in 1953 (Dow Chemical Company), with sandwich baggies following a few years later. Sealable zipper baggies were introduced for commercial use in 1959 (Minigrip). Since then, there have been numerous innovations in the design and use of sealable plastic baggies. Options now include biodegradable bags, bubble-wrap bags, antistatic bags for shipping electronic parts, bags for containing medical specimens, as well as bags for preserving breast milk and many other purposes.

The zipper was invented in 1851 and originally called an "automatic continuous closing closure" (Smithsonian Libraries 2010). Zippers did not become widespread for many decades, until they were used in rubber boots in the 1920s. The name "zipper" caught on at this time, as well, as it resembled the sound made when the device was opened and closed. It was not until the late 1930s that French designers began using zippers in men's pants and other clothing. In either the Engage or the Extend portion of the lesson, you may wish to have students do some of their own research related to the history and development of plastics for household uses (for a start, see References).

FIGURE 7.1 Sealing mechanism of two double-track baggies: (a) open, (b) closed

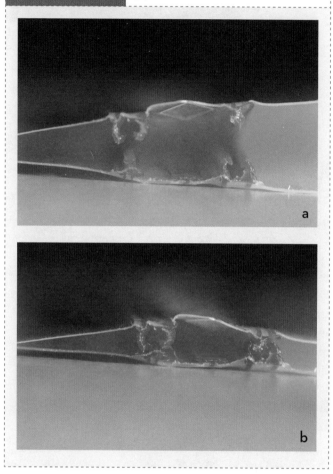

FIGURE 7.2 Sealing mechanism of two multitrack baggies: (a) open, (b) closed

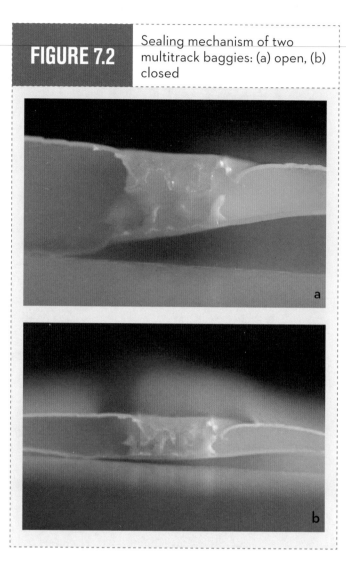

Investigating Sealable Bags (Teacher Background Information)

Materials

For this lesson, you will need to gather several different brands of sealable baggies (three or four per group of students). Sandwich or snack size will each work well, as we are focusing only on the sealing mechanism. Because it is easier to observe the sealing mechanism if the seal is colored, check the box to see if the seal is colored or not. You may also wish to cut each baggie in half through the

seal so students can view the end of the sealing mechanism (Figure 7.1). For ease of discussion, you might want to label the bags distributed to students. For the Extend stage of the lesson, you will need to have sealable baggies of the sliding zipper type. Sandwich-size baggies range in price from 3 to 15 cents each, depending on brand and style. Ideally, each student will also have a hand lens.

Engage

Initiate a discussion about how students' families pack sandwiches or other lunch or picnic foods. Some may

FIGURE 7.3 Sealing mechanism for folders: (a) with straws, (b) with straw and pipe cleaner

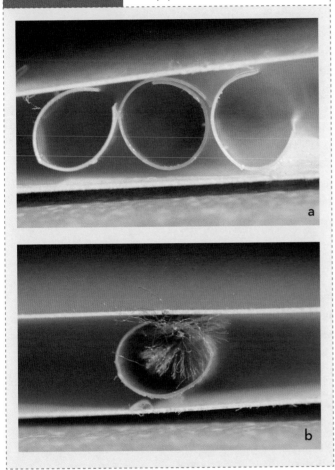

FIGURE 7.4 Slider on sealable baggies: (a) with protrusion for separating, (b) for sealing

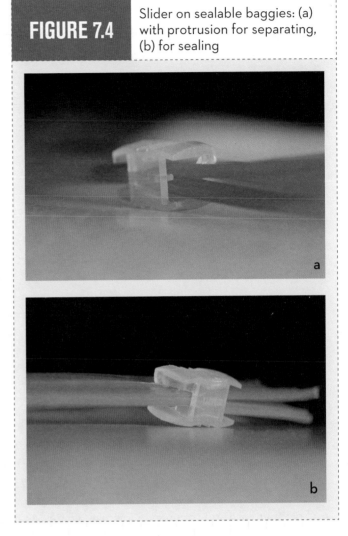

use hard plastic reusable containers, while most others likely use disposable plastic wrap, plastic baggies, or foil (you will probably find that most of your students' families use the sealable plastic baggies).

Distribute at least three different types of sealable plastic baggies to each group of two to four students and have them look at the sealing mechanism with a hand lens of any magnification. Instruct students to make and label a sketch that explains how each of the mechanisms works. The basic design is some type of a rib that fits snugly into a slot. The ribs and the slots often have slight hooks that grab hold of each other.

This is referred to as a single-track type of seal. There are also mechanisms that seal by means of two or more separate tracks—two ribs that fit into two separate slots, known as double-track designs (Figures 7.1a and 7.1b). Finally, other baggie seals have a multitrack that consists of at least two ribs that fit into one or more slots (Figures 7.2a and 7.2b). Students will likely observe numerous variations on these three basic mechanism designs. As an optional extension, students can consider forces (where it is separating, squeezing, rubbing, sliding, etc.).

FIGURE 7.5   Zipper teeth

Explore

Safety note: Students should wear safety goggles for this entire exploration. Review safe use of the low-temperature hot glue gun and scissors (see Roy 2010 for glue gun safety guidelines).

Now that students have observed the various ways bags are sealed, they will be designing their own mechanism to "seal" two sides of a file folder together (Figure 7.3, p. 51). For each group of two to four students, provide a file folder, small file folder scraps for preliminary testing, scissors, a low-temperature glue gun, and a variety of materials to make ribs and slots (e.g., pipe cleaners, straws, coffee stirrers, etc.).

Students should first make a sketch and seek your approval before they begin to construct their models. This will allow you to be aware of needed materials and to ensure students are following safe laboratory procedures. Have students use the scraps of folder to test their designs before finally attaching the sealing mechanism to the file folder. Most students should be able to do this within one class period. If not, each group can store projects from one class period to the next in a plastic tub, shoe box, or large envelope.

Explain

Have students write an explanation of how their sealing mechanism works, including arrows indicating where the forces are acting to hold the pieces together. In general, the rod fits into the slot, and either the rod forces the slot apart or the slot squeezes the rod so that adequate friction holds the two together. The rod is a bit bigger than the initial opening of the slot and pushes it apart as it is pushed in. The slot springs back after the rod is inserted to hold it in place. Friction between the slot and the rod helps to keep it in place, as well. To open the mechanism, one must overcome this force of friction as well as the force needed to push apart the two sides of the slot. The force required to pull the two locking members of the seal apart is known as the *holding force* (Fresh-Lock Zipper). Students should also note whether their design is single track, double track, or multitrack. Have students share their sealing mechanisms for the file folders with the class. Encourage students to analyze how each group's mechanism works. Have a discussion about how the various designs are similar and different—both compared to each other and to the designs used on the plastic baggies.

Extend

For this stage, each student group will need a zipper-type baggie, a hand lens, and one metal zipper. Metal zippers of the type used for jeans can be purchased for approximately $1.50 each or can be recycled from old clothing. The bottom of the zippers must be cut, but they can be put back together to be used from class to class. The teeth on most plastic zippers are too small to be seen easily. Cut through the teeth of the zipper just above the bottom stop so that the slider can be removed.

The slider-type seals on the plastic baggies work essentially in the same way as the nonslider types. They will all have a rib and a slot (or several) that fit together. Look closely and note that one end of the slider has a protrusion that separates the two halves of the baggie when pulled in one direction (Figure 7.4a, p. 51). When slid in the opposite direction, the other end of the slider pushes the two sides together in the same way fingers do when sealing a nonslider baggie (Figure 7.4b, p. 51). Have students locate the protrusion on their slider and see how it is used to separate the two halves.

Distribute a zipper and hand lens to each group of students. Students will investigate how the zipper works, first by moving the slider up and down and observing how the teeth mesh together. After they remove the slider from the cut zipper, have them try to mesh the two sides together. This can usually be done by holding the two sides at a wide angle and, starting at the bottom, meshing the two sides together. The teeth on most metal zippers have a point on the top and an indentation on the bottom that interlock (Figure 7.5). Unlike the sealable bags, the teeth on both sides of the zipper are the same, only slightly offset.

Other ways to extend the lesson might be to compare the cost of wrapping a sandwich in wax paper, aluminum foil, plastic wrap, and sealable baggies. What are the advantages and disadvantages of each method? Students might also investigate other sealing mechanisms such as Velcro, and compare it to how burrs stick to the fur of animals.

Evaluate

To give students further opportunities to use their creativity to generate innovations, have them brainstorm uses for the zipper, the sealable mechanism, or their own design from the exploration. You may wish to have students conduct a home or school scavenger hunt to identify as many uses for sealable bags as they can. For example, the original use of sealable plastic fasteners was for school pencil bags. Students should select one idea—the zipper, sealable plastic fastener, or their own sealing design—and explain a new use for that mechanism.

Conclusion

Engineering is often a very creative process, and engineers must use this creativity to be problem solvers. The zipper existed for nearly 100 years before a variation, the sealable Ziploc plastic baggie, was developed that made use of a zipperlike slider but with a different connecting mechanism. Both zippers and sealable plastic fasteners are a part of our everyday lives. We no longer zip just our boots and sandwich bags; with a little creativity, who knows what we will be zipping in the future.

References

American Chemistry Council. The history of plastic. *www.americanchemistry.com/s_plastics/doc.asp?CID=1102&DID=4665*.

Dow Chemical Company. Our company—History: 1950s. *www.dow.com/about/aboutdow/history/1950s.htm*.

Fresh-Lock Zipper. Resources—Presented by Fresh-Lock Zipper. *www.fresh-lock.com/resources*.

International Technology Education Association (ITEA). 2002. *Standards for technological literacy: Content for the study of technology.* 2nd ed. Reston, VA: ITEA.

Minigrip. About Us—Minigrip Commercial. *www.minigrip.com/about.html*.

Roy, K. 2010. Glue guns: Aiming for safety. *Science Scope* 34 (2): 84–85.

Smithsonian Libraries. 2010. The up and down history of the zipper. *http://smithsonianlibraries.si.edu/smithsonianlibraries/2010/05/the-up-an-down-history-of-the-zipper-.html*.

Resource

Moyer, R. H., J. K. Hackett, and S. A. Everett. 2007. *Teaching science as investigations: Modeling inquiry through learning cycle lessons.* Upper Saddle River, NJ: Pearson/Merrill/Prentice Hall.

ACTIVITY WORKSHEET 7.1 Investigating Sealable Bags

Engage

1. Some people wrap a sandwich in plastic wrap or aluminum foil or put it in a baggie. How do you pack a sandwich for your lunch?

2. Look at the sealable baggies your teacher has provided and open and close each of them several times. Now use a hand lens to take a closer look at the sealing mechanisms. Notice how they are similar to and different from each other.

3. Make a drawing of at least two different sealing mechanisms. Write an explanation of how the sealing mechanism works. Share your thinking about the mechanisms with your classmates and your teacher.

Explore

Safety note: Wear safety goggles during this exploration.

1. Design your own sealing mechanism that will hold a file folder closed. Make a sketch of your plan.

2. After your teacher has approved your plan, construct a working model using the materials that have been provided.

3. You may wish to modify your design after you have tested it. Finally, attach your best model to the file folder.

Explain

1. Write an explanation of how your mechanism holds the two sides of the folder together. Include a description of the forces that hold the two pieces together. Make a drawing that includes arrows to indicate the forces.

2. Compare your design to the sealing mechanisms of the baggies you observed. Write a description of how yours is similar to and different from the baggies. Have you made a single, double, or multitrack sealing mechanism?

3. Share your models and explanations with the class and discuss the different methods used to seal the folders together.

Extend

1. Look at a sealable baggie that has a sliding zipper-like device. How is the sealing mechanism similar to and different from the baggies you have already observed?

2. What does the slider do as you move it back and forth? Record your observations and discuss your understandings with the class.

3. Look at the sample zipper. Zip and unzip it a few times and see if you can determine how it works. Remove the zipper pull from the cut zipper that your teacher provided. See if you can mesh the teeth of the zipper together without the pull.

4. Make a sketch of how a zipper works. Record how it is similar to and different from the zipper-type baggie.

Evaluate

Select one idea—the zipper, the sealable plastic fastener, or your own sealing design from the Explore section—and invent a new use for that mechanism. Write an explanation of your innovation.

PART 4

Bathroom Engineering

CHAPTER 8

AN ABSORBING LOOK AT TERRY CLOTH TOWELS

WHEN YOU DRY off after a shower, you may not realize that you have used a piece of technology. But the common bath towel is the result of everyday engineering. The textile industry employs engineers and scientists to design new fabrics and to find economical ways to manufacture them. Examples range from moisture-wicking fabrics for sportswear, to wrinkle-resistant shirts and slacks, to outdoor materials that are less susceptible to mildew and fading from the Sun.

In this lesson, students explore the absorbency of several towels with different weaves and weights. The lesson follows the 5E Model and incorporates engineering in the sense of product testing with a focus on the relationship between the weave of a towel and its absorbency. The National Science Education Standards indicate that middle-level students should understand "a substance has characteristic properties, such as density, boiling point, and solubility" (NRC 1996, p. 154).

Historical Information

The common terry cloth bath towels with which we are familiar have only been used in the Western world since 1840, when Henry Christy, a British business-

man and explorer, brought them back from Bursa, Turkey. Turkey was a stop on the Silk Road, a trade routed followed by Europeans on their way to China. Turkish towels are still considered some of the finest in the world.

Look closely at a terry cloth towel (Figure 8.1a and 8.1b, p. 58) and you will notice a series of loops. These loops increase the towel's total surface area and increase the towel's ability to absorb moisture. Cotton is able to absorb seven to eight times its weight in water. Synthetic microfibers, because of their extremely fine threads, are also able to readily absorb water (Figure 8.1c). By increasing the amount of material in the towel via the weave, the absorbency can vary. From an engineering point of view, the design of towels has evolved over time. As noted in the *Standards for Technological Literacy*, "the specialization of function has been at the heart of many technological improvements" (ITEA 2002, p. 83). Prior to the Turks inventing the loop pile we know as terry cloth, towels were much less able to absorb water. We now have a wide assortment of towels for different purposes. Towels come in different sizes—from washcloths to beach towels—and are made of different materials and weaves resulting in varying levels of absorbency, lint, and softness. For example, linen tea towels used for lint-free drying of glassware

CHAPTER 8

FIGURE 8.1

Close-up (10×) of four towels: (a) terry cloth, (b) longer looped terry cloth, (c) microfiber loops, and (d) flat-weave dishcloth

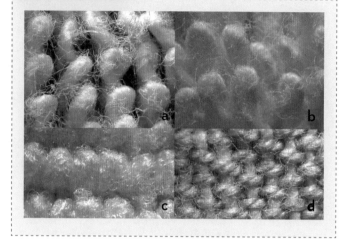

group. A standard bath towel is about 68.6 cm × 132 cm (27 in. × 52 in.). You can make 28 samples (17.1 cm [6.75 in.] per side) from each bath towel. Other options would be to purchase samples at a fabric store or have students bring in washcloths or dishcloths from home ahead of time. Slightly smaller bath towels can be purchased online for less than $2 apiece. Distribute at least three different towel samples and hand lenses to each group of three to four students and direct them to observe the weaves of each towel. Students may look at the towels with a stereoscope if available; a standard magnifying glass will work, as well. Samples may include typical looped terry cloth, microfiber, cut pile, or simply flat-weave fabrics. Ask students to predict which towel they think will absorb the most water based on their observations.

Explore

Safety note: Students should wear safety glasses.

In addition to the towel samples, each group will need a beaker (400 ml or larger—any jar or container can be used as long as each group uses the same size to allow for comparison among groups), pipette, rubber band large enough to fit around the beaker, and a container of water. You should have a few extra towel samples, because once they are wet, they cannot be used again until completely dry. You will need separate sets of towels for multiple classes.

Students should secure a towel sample over a beaker with a rubber band as shown in Figure 8.2. Although we have described a procedure in Activity Worksheet 8.1, you may wish to have your students determine their own method for testing the towels' absorbencies. You can discuss which variables need to be measured and controlled. In our procedure, there needs to be a small depression in the towel so that the water will eventually drip into the beaker rather than over the sides. Students add water one pipette at a time until the towel can no longer absorb any more water, and it begins to drip into the beaker. Students should record this amount of water. Many pipettes have a

do not have a loop pile, but are made from a flat weave similar to a dishcloth (Figure 8.1d).

Investigating Towel Absorbency (Teacher Background Information)

Engage

While displaying at least two different towels (one obviously plusher than the other), initiate a discussion about what students like in a towel and which of the two on display they would choose to dry themselves. Have students brainstorm and share characteristics of towels they have used. You may also wish to introduce some of the history of towels at this time and ask students to point out Turkey on a world map. Perhaps some students are familiar with the area and the culture.

You will need samples of several different types of towels for this exploration. You could use washcloths and dishcloths or cut larger towels into small pieces (large enough to cover at least a 400 ml beaker). To complete all parts of the lesson, you will need to have five dry samples of various types of towels for each

FIGURE 8.2 Testing setup

FIGURE 8.3 Volume of water absorbed

Type of towel	Water held (ml)
Microfiber	11
Plush terry cloth towel	30
Tea towel	1.1

volume of about 1 ml. This can be calibrated with a graduated cylinder. Students can construct their own data table, or you can provide one similar to Figure 8.3 that includes sample results. Students then repeat this procedure for each of their towel samples.

Explain

Have students share their results. Because students were only able to do one trial, the class data can be considered to represent multiple trials. While the class results will likely vary, the order of the towel absorbency should be the same for each group. Several variables that will affect a towel's absorbency should arise from the discussion of their results—the type of material, the amount of material (or density), and a related factor, the weave. Cotton is more absorbent than most synthetic fibers and cotton/polyester blends. As noted above, cotton can absorb up to seven or eight times its mass in water. Thus, if more cotton can be woven into the same area, the towel's absorbency increases. Towel designers measure a towel's grams per square meter (g/m²), or GSM, a measure of the mass of cotton per a given area. A towel's GSM can be increased by adding loops at a 90° angle to the basic weave, which creates typical terry cloth. The GSM can also

be increased by making the loops out of finer threads, such as Egyptian cotton, which results in more threads in the yarn that makes up a loop (Figure 8.4, p.60). In other towels, the loops are made longer, as seen in Figure 8.1. Some plush towels are made to feel soft by cutting the loops (Figure 8.5, p. 60) on one side, but that actually decreases the GSM of the towel—it feels softer, but it absorbs less water (Youso 2007). Some Turkish towels have a GSM of more than 700, while a T-shirt might be about 150 GSM. Provide students with a ruler so they can calculate the GSM of one or more of their samples. A typical terry cloth washcloth had the following measurement:

$$Area = 0.31 \text{ m} \times 0.28 \text{ m} = 0.09 \text{ m}^2$$
$$Its \text{ mass is } 40.4 \text{ g. Therefore,}$$
$$GSM = 40.4 \text{ g}/0.09\text{m}^2 = 449 \text{ g/m}^2.$$

Have students compare their results with other groups' to determine if absorbency increases with GSM (it should). Be sure that students only make GSM comparisons between cotton towels or between microfiber towels for consistency.

Extend

In this section, students compare the mass of wet towels to dry towels as another measure of absorbency. Provide each group with at least one more dry towel sample. If possible, give each group a different type of towel. Students need to mass the dry towel, and then

FIGURE 8.4 Close-up (40×) of a cotton terry cloth loop

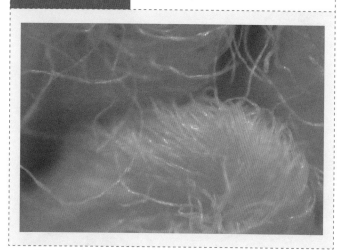

FIGURE 8.5 Close-up (10×) of a cut-pile towel

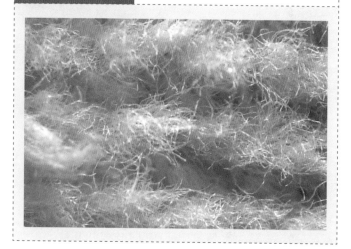

soak it in a beaker of water. As suggested in Activity Worksheet 8.1 (p. 62), you can have students "standardize" the amount of wetness by letting each sample drip for a minute. However, you may want to have a discussion with the class regarding their ideas for such standardization of procedure, as well as its importance. We recommend using a baggie to protect electronic balances (Figure 8.6). Students find the mass of the water that the towel absorbs by subtracting the mass of the dry towel and the baggie from the mass of the wet towel in the baggie. Students should find that their samples of terry cloth will hold somewhere between three to eight times their mass in water. Dishcloths and tea towels will hold significantly less. Microfiber towels are made from polyester and nylon using extremely thin (about 100 times thinner than a human hair) threads (Figure 8.1c, p. 58) that greatly increase the towels' GSM and ability to absorb water. See Figure 8.7 for sample results.

Evaluate

Have students explain in writing the difference in drying ability between a thin, worn-out towel and a new, plush one. They should note that the newer towel has more loops and a greater GSM, thus more surface to absorb water.

Conclusion

While it may seem that little has changed in terry cloth towels over the last 150 years, they have indeed evolved in numerous ways. Microfiber towels are becoming extremely popular due to their softness, absorbency, and cleaning ability (the tiny fibers are able to trap dirt). Fabric engineers have also developed other new types of fabrics and ways to weave them to address other needs. Engineering design is driven by the "specialization of function." Modern fabrics, for example, have been developed that are able not only to wick moisture away from an athlete's body but also to offer adequate insulation against the cold. You could ask students to research some of these developments, make a timeline of how fabrics have changed over time, and report the timeline to the class. Students may be able to obtain some older clothing and fabrics from senior relatives. Middle-level students should appreciate the wide array of engineering and design applications in everyday life.

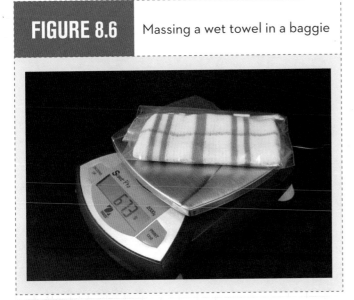

FIGURE 8.6 Massing a wet towel in a baggie

References

International Technology Education Association (ITEA). 2002. *Standards for technological literacy: Content for the study of technology.* 2nd ed. Reston, VA: ITEA.

Moyer, R. H., J. K. Hackett, and S. A. Everett. 2007. *Teaching science as investigations: Modeling inquiry through learning cycle lessons.* Upper Saddle River, NJ: Prentice Hall.

National Research Council (NRC). 1996. *National science education standards.* Washington, DC: National Academies Press.

Youso, K. 2007. Fixit: Loopier the weave, the better the towel. *Minneapolis Star Tribune.* www.startribune.com/lifestyle/homegarden/11311456.html.

FIGURE 8.7 Mass of water absorbed

Type of towel	Dry mass of towel (g)	Baggie mass (g)	Mass of wet towel and baggie (g)	Mass of water absorbed (g)	Mass of water/mass of towel (g)
Dish cloth	19.0	3.4	67.3	44.9	2.4
Microfiber	28.1	3.5	165.5	133.9	4.8
Terry cloth	40.4	3.5	184.2	140.3	3.5

ACTIVITY WORKSHEET 8.1   Investigating Towel Absorbency

In this activity, you will examine a number of towels and determine if their properties affect the amount of water that they can absorb.

Engage

1. Examine the towels provided by your teacher. Which would you rather use to dry off after swimming or a shower?
2. Based on what characteristics of the towel did you make your choice? Record your ideas and share with your classmates.
3. Obtain the towel samples from your teacher and note the characteristics of each. Use a hand lens to study the weaves of the towels. Draw and label what you see.
4. Based on your observations, which sample will be able to absorb the most water? Give reasons for making this prediction.

Explore

Safety note: Wear safety glasses to protect your eyes.

1. Attach one of the sample towels over the top of a beaker with a rubber band. Make sure there is a small indentation in the top of the towel so that water can drip into the beaker.
2. Use a pipette to add water to the depression in the towel sample on top of the beaker. Count the number of whole pipettes of water you add before the towel can no longer hold any more and some of the water drips into the beaker. Make a data table and record your data.
3. Repeat steps 1 and 2 for each of your towel samples.

Explain

1. Share your results with your classmates. Which towel absorbed the most water? What steps in

the procedure could have caused errors or flaws in the data?
2. Rank the towels from least to most absorbent. Describe how the characteristics of the towels change over this range. How do the weaves differ?
3. Determine the grams per square per meter (GSM) of a dry cotton towel sample by dividing its mass in grams by its area in square meters. How does the GSM compare to the amount of water absorbed?
4. How do you think your rankings compare to the cost of the towels? When might people choose to use towels of different absorbency levels?

Extend

1. How does the mass of water a towel can absorb compare to the dry mass of the towel?
2. Obtain one more dry towel sample. Mass and record. Mass an empty baggie and record.
3. Immerse the towel in a beaker of water until it is saturated. Remove the towel and let it drip for a minute and then place it in the baggie and mass once again.
4. How can you find the mass of the water that the towel has absorbed?
5. Can a towel absorb a mass of water that is more than the towel? How do your results compare with your classmates'?

Evaluate

You have a wet pet. Your parent has given you an old, thin, worn-out towel. You want to use a newer, plusher towel. How can you convince your parent that the newer towel will dry your pet better? Explain your thinking in writing and include a drawing.

CHAPTER 9

TOOTHBRUSH DESIGN— IS THERE A BETTER BRISTLE?

WHAT KIND OF toothbrush do you use—manual or electric? What is the shape of the head and the handle? Could you describe the firmness and the layout of the bristles? Or count the number of bristles? With more than 3,000 toothbrush patents, is one type better than others for cleaning your teeth? Manufacturers often claim that their particular design is better than the competitors, but is it?

In this 5E Model lesson, students explore various manual toothbrush designs. One of ITEEA's standards urges students in grades 6 through 8 to learn that "there is no perfect design" (ITEA 2002, p. 95). As in the design of all products, engineers are faced with prioritizing costs and benefits. For example, a very simple toothbrush (see left-hand toothbrush in Figure 9.1, p. 64) can be purchased for 50 cents. A more complex toothbrush that has variable bristles intended to clean between different tooth surfaces and a bent handle to allow easy access to the hard-to-reach back of the mouth may cost $3–$5. Engineers must consider the economic issues as well as the functionality involved with selling the products they create: to produce the best possible toothbrush regardless of cost, the toothbrush that will sell the most, or perhaps the toothbrush that costs the least to produce.

As with other lessons in this book, there is a connection to science content, as well. It is important for students to appreciate the value of proper dental hygiene. Regular effective brushing protects the body from dental caries (i.e., tooth decay and cavities) and periodontal disease, which may be associated with other health concerns, including heart health. You can use this lesson when focusing on the following National Science Education Standard: "Disease is a breakdown in structures or functions of an organism" (NRC 1996, p. 157).

Historical Information

Primitive toothbrushes have been found in tombs of ancient Egyptians, evidence that humans have been using devices to clean their teeth for at least 5,500 years. These first tooth "brushes" were usually made of twigs that had been frayed at one end to clean between teeth. In the 15th century, the Chinese used boar bristles stuck into a bone or a stick—sometimes made from aromatic trees—to brush their teeth, as well as to freshen the breath. Two hundred years later, Europeans were using rags soaked in salt solutions to clean their teeth. In 1780, William Addis made a toothbrush using hairs from a cow's tail attached to a handle carved from an ox's thighbone. The Addis family is still producing toothbrushes in England today. Toothbrush design did not change all that much until 1938, when nylon toothbrushes were introduced in the United States. Nylon proved to be beneficial because the bristles could

FIGURE 9.1 Toothbrush designs

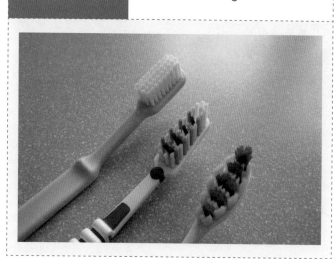

FIGURE 9.2 Bristles at 30×

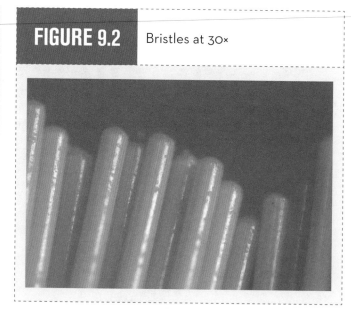

be shaped on the ends not only to be gentler but also to reach spaces between teeth and under the gum line (Figure 9.2). It also proved to be more hygienic, because the bristles were more resistant to bacteria growth than animal-hair bristles. It was not until after soldiers came home from World War II in 1945, however, that the idea of brushing one's teeth became popular with most Americans (ADA 2007).

During this activity, students may logically inquire about the purpose of toothpaste. You may want to have them research this on their own. Essentially, toothpaste serves a number of purposes. Most toothpastes contain a detergent to help clean the teeth. A flavor enhancer masks the taste of the detergent and freshens the breath. Most, but not all, toothpastes contain a fluoride compound to help prevent tooth decay. Children's toothpaste contains about a third less fluoride because they are more likely to swallow it. Toothpastes also usually contain some abrasive materials—a fine grit—to help scrub away plaque, and whiteners to remove superficial stains. Some people prefer natural toothpaste alternatives made of organic herbal material.

The earliest references to toothpastes date to 5000 BC in Egypt, where a paste was described for cleaning teeth. Over the years, toothpaste has evolved and has included various abrasives—some actually harmful to the enamel of the teeth. Before about 1850, tooth cleaners were actually powders and not pastes. The first toothpaste tube was introduced in 1890, and tube containers—another example of everyday engineering—are still used today (Colgate World of Care 2006).

Investigating Toothbrushes (Teacher Background Information)

Engage

Safety note: Remind students that marshmallow cream may not be eaten in the lab.

You will need to gather an assortment of toothbrushes of different designs. You might consider contacting local dentists for donations or purchasing some at your local dollar store. A less desirable option is to collect an assortment of used toothbrushes and then sterilize them by boiling or bleaching. For the Explore stage, each group of three to four students will need two different toothbrushes and two additional, more

FIGURE 9.3   Exploration setup

FIGURE 9.4   A more complex toothbrush

complex toothbrushes for the Extend phase. After rinsing, the toothbrushes can be reused for later classes. In addition, each group will need two plastic combs, two disposable cups, a cup of water to rinse brushes, a plastic knife, about 60 cm of electrical tape, and about 15 ml (1 tbsp.) of marshmallow cream (some groups may request additional cream for the second part of the Exploration). Also, students will need a few general supplies: some kind of tape (duct tape or electrical tape) to hold the materials to the lab table, scissors, and newspaper to protect the floor. Because students are planning their own experiments, they may request additional materials.

Even though we all use our toothbrushes (hopefully) at least twice a day, it is likely that most of us have never really studied this simple example of everyday engineering. Therefore, before distributing the toothbrushes, ask students what they know about toothbrushes and dental hygiene in general. You may wish to have students investigate some of the history of toothbrushes. Next, have students consider what their own toothbrushes look like and ask them to make a sketch. It is likely that students will have difficulty recalling specific details (e.g., whether the bristles are

all the same height, the number of tufts in a row, the direction the tufts point, or the number of rows of tufts). After students share their drawings, distribute two different brushes to each group and ask students to compare them. Help students realize that toothbrushes have a range of features. In the second part of the exploration, students will test to discover if any of these differences result in a more effective toothbrush.

Explore

In this activity, students will set up two models of a mouth and teeth using a disposable cup and a plastic comb as shown in Figure 9.3. Students should place a piece of electrical tape around the base of the comb's teeth to represent the gum line and then evenly smear marshmallow cream on the teeth to simulate plaque and food particles. Each group of students should plan a fair test to determine whether or not one of their toothbrushes is more effective at cleaning the marshmallow cream off of the teeth of the combs.

One possible procedure might be to brush five strokes in one direction on each side of the comb, rinsing in between sides. To operationalize the dependent variable of effectiveness, students can visually compare

FIGURE 9.5 Toothbrush design features

Design feature	Purpose
Bent handle	Cleans hard-to-reach teeth in back of mouth
Thumb grip	Properly positions brush in hand to reach all parts of mouth
Flexible handle	Reduces pressure on gums
Soft rubber bristles along edge	Clean and massage along gum line
Bristles pointing in different directions	Clean spaces between teeth
Bristles of different lengths	Clean spaces between teeth
Rough area on back of brush	Scrubs the tongue and inside of the cheeks

the amount of marshmallow cream remaining on each comb. Students can also compare the amount remaining underneath the gum line.

In the second part of the exploration, students will determine how to get the model teeth as clean as possible. Here, the driving question is whether it is possible to clean the teeth regardless of which brush is used. Students will need to reapply the marshmallow cream and try out different brushing techniques. Students should make note of the techniques used.

Explain

Have students share their findings and compare differences and similarities. It is likely that students will be unable to see significant differences between their two toothbrushes if the techniques are carefully controlled. Students may be surprised that their results show little or no difference between expensive and inexpensive toothbrushes. Most of the toothbrushes used in this activity (especially if they came from dentists or major retail outlets) were probably approved by the American Dental Association (ADA). The ADA evaluates

toothbrushes and gives the ADA Seal of Acceptance to those that meet their requirements. Currently, the approved list includes 34 different toothbrushes (ADA 2010); therefore, all should clean the teeth effectively. In fact, any toothbrush is likely to clean the model teeth similarly.

With sufficient brushing, students should be able to fully clean the model teeth with both of their brushes. Studies have shown that "there is no convincing evidence to support the idea that one type [of toothbrush] is better than the other in terms of its efficacy in plaque removal" (Sasan et al. 2006, p. 168). Students should conclude that the key factor is to brush for a sufficient amount of time and to reach all surfaces, including in between the teeth and along the gum line. To help answer additional questions, you may wish to ask a local dentist or hygienist to visit your class.

Extend

Provide two additional toothbrushes with more complex features (Figure 9.4, p. 65) to each group. These toothbrushes might have a curved or flexible handle,

FIGURE 9.6

FIGURE 9.6 | Sample scoring rubric for toothbrush design activity

	3	2	1
Design	Creative design clearly meets some specific purpose.	Purpose and design are not well matched.	Purpose or design is not clearly stated.
Toothbrush information	Accurate information about toothbrush characteristics and functions is provided.	Some information about the toothbrush design is provided.	Inaccurate information about toothbrush characteristics and functions is provided.
Communication	Creatively and clearly communicates the selling points of the toothbrush.	Clear but not creative communication of the selling points is provided.	Communication of selling points is lacking or ineffective.

a thumb grip to properly position the brush in the hand, different types of bristles, and so on. Students should study the features of their brushes and make inferences as to their purpose. See Figure 9.5 for sample observations and inferences.

Evaluate

Because a perfect design does not exist, students should articulate some specific characteristics that a person may value in a toothbrush by creating an advertisement. For example, if a student values a very inexpensive toothbrush, the advertisement could indicate that people should take the time and effort to brush all tooth surfaces effectively. Students could either design a print advertisement or use a video camera to produce a video commercial. Students can make a drawing of their design rather than making the actual toothbrush itself. See Figure 9.6 for a sample scoring rubric.

Conclusion

The toothbrush analysis is a good example of the ITEEA standard that "there is no [one] perfect design"

(ITEA 2002, p. 95). There are many equally effective toothbrushes from which to choose. As the ADA states, "Choose a toothbrush that you like and find easy to use so that you'll use it twice a day to thoroughly clean all of your tooth surfaces" (2007, p. 1288). Students should appreciate that in some cases a less complex engineering solution may be effective if properly used or applied. Toothbrush design is but one factor in effective oral hygiene, which is also impacted by brushing time, technique, and frequency.

References

American Dental Association (ADA). 2007. A look at toothbrushes. *Journal of American Dental Association* 138: 1288.

ADA. 2010. ADA Seal of Acceptance program and products. *www.ada.org/sealprogramproducts.aspx.*

Colgate World of Care. 2006. History of toothbrushes and toothpastes. *www.colgate. com/app/Colgate/US/OC/Information/ OralHealthBasics/GoodOralHygiene/Brush ingandFlossing/HistoryToothbrushesToothpastes.cvsp.*

International Technology Education Association (ITEA). 2002. *Standards for technological literacy: Content for the study of technology.* 2nd ed. Reston, VA: ITEA.

Moyer, R., J. Hackett, and S. Everett. 2007. *Teaching science as investigations: Modeling inquiry through learning cycle lessons.* Upper Saddle River, NJ: Pearson/Merrill/Prentice Hall.

National Research Council (NRC). 1996. *National science education standards.* Washington, DC: National Academies Press.

Sasan, D., B. Thomas, M. Bhat, K. S. Aithal, and P. R. Ramesh. 2006. Toothbrush selection: A dilemma? *Indian Journal of Dental Research* 17 (4): 167–70.

ACTIVITY WORKSHEET 9.1 — Investigating Toothbrushes: Toothbrush Design—Does It Matter?

Engage

Safety note: Do not put toothbrushes or any of the lab materials in your mouth.

1. Make a drawing of what your toothbrush looks like. Try to remember how it is shaped and what the bristles look like.
2. Compare the two toothbrushes your teacher has provided. Note how they are similar and how they are different.
3. In this activity, you will design and conduct a test to determine if one of these brushes is more effective than the other. Write your prediction as well as your reasoning and discuss with your group.
4. Can you find a brushing technique that will allow you to clean teeth effectively with either brush?

Explore

Safety note: Marshmallow cream used to simulate plaque and food particles may not be eaten.

1. Discuss with your group a plan to conduct a fair test of the two toothbrushes using the model teeth and mouth your teacher has supplied.
2. In your plan be sure to include the following:
   a. A set procedure for brushing—number of strokes, pressure used, direction of brushing, rinsing of brush, and so on
   b. How you will determine the effectiveness of the toothbrushes
3. After your teacher has approved your plan, set up and conduct your test.
4. Record your findings.
5. Using the same materials, investigate what you have to do to get your model teeth as clean as possible. Compare different brushing techniques. Record your findings.

Explain

1. Share your group's findings with the rest of the class. How do your findings compare?
2. Was any one brush obviously more effective than others? What conclusion might you draw from this part of the investigation?
3. Discuss with the class the various methods used to get the model teeth as clean as possible. How well did each brush clean the model teeth?

Extend

1. Examine the different toothbrushes your teacher has provided.
2. Notice the variation in the design of each brush.
3. Thinking like an engineer, what do you think is the intended purpose for each design feature of each brush?

Toothbrush	Design feature	Purpose of feature
One		
Two		
Three		
Four		

4. Present your ideas to the rest of the class and discuss.

Evaluate

Review some advertisements for toothbrushes to see what different types exist. What are the advantages that the commercials stress as selling points? Create an advertisement for a new toothbrush noting the purpose and advantages of your design.

PART 5

Electrical Engineering

CHAPTER 10

HOLIDAY BLINKERS

MANY EXAMPLES OF engineering go unnoticed because they are so much a part of our daily lives that we rarely give them much thought. When you decorate your house with holiday lights, you probably do not think about the engineering that was needed to produce them. Holiday lights have become safer, cooler, and more economical to produce and operate since their invention more than 100 years ago. Here we will examine how current flowing through a bimetallic strip causes lights to blink on and off. A bimetallic strip is made of two different thin metals, usually copper (or brass) and steel that are bonded together. The steel expands less than the copper or the brass when heated, so the strip curves toward the steel side.

Historical Information

The first electric holiday tree lights were made in 1882 by Edward Johnson, an employee of Thomas Edison. Johnson wanted a safer way than candles to decorate his tree. The idea was slow to catch on, and it was not until after 1895 when President Cleveland had electric lights on the White House family tree that holiday lights became popular with the public. The first lights were extremely expensive and only the wealthy could afford them. Around 1950, the miniature lights that were the precursors to our modern lights were introduced in Italy (Nelson 2008). These lights were small and connected in series so that they were cooler and less expensive to operate than the larger parallel wired lights popular at that time. While series holiday lights were popular prior to the 1950s, they were larger and had fewer lights on a string—often eight. Today, most miniature lights contain two 50-bulb strings parallel with one another to make the common 100-bulb sets.

The first twinkling lamps were the larger parallel types, and each bulb flashed on and off individually. The most common inexpensive blinking lights in use today are the miniature variety in series, in which one bulb blinks, causing the remaining bulbs to flash on and off. The newest blinking lights make use of electronics that can be programmed to create numerous intricate designs and flashing patterns.

The short history of holiday lights demonstrates a key idea in engineering—that engineering is not simply the invention of an idea, but also the idea's evolution through innovation: "Invention is the process of turning ideas and imagination into devices and systems. Innovation is the process of modifying an existing product or system to improve it" (ITEA 2002, p. 110). After the initial invention of holiday lights on a string, many people developed innovations to improve upon them.

FIGURE 10.1 Comparison of regular (on left) and blinker bulbs (on right)

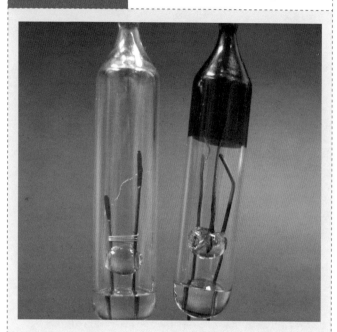

In this lesson, like in all Everyday Engineering lessons, we will integrate ITEEA standards (like the one on page 73) for grades 6–8 with appropriate science content. We will present a 5E Model lesson: Making a switch from a bimetallic strip to create a blinking bulb.

Investigating Blinking Bulbs (Teacher Background Information)

Engage

Show students a string of miniature holiday lights (available at holiday time at most hardware stores for about $2). Demonstrate what happens when you replace one of the bulbs with a blinker or flasher bulb (see right bulb in Figure 10.1). Students will note that all of the lights blink on and off. Have students share their ideas about how the blinker bulb caused the rest of the lights to go on and off. If necessary, review at this time what students already know about series circuits (see Series and Parallel Circuits). Explain to students that they are to observe a blinker bulb and a regular bulb in order to determine how they are different and make drawings of the two types of bulbs.

Explore

Safety note: Do not use rechargeable cells because they can get very warm if students accidentally create a short circuit.

You will need to cut apart an old string of holiday lights to have individual segments with a bulb and 8–10 cm of wire on both sides so that the ends of the wire will reach each end of a battery. The ends of the wire will need to be stripped. Each group of two to four students should have one segment with a regular bulb and one with a blinker bulb, two D-cell batteries, and some pieces of electrical tape (Figure 10.2). Students will note that the blinker goes on and off. Their drawings of the regular bulb should include two posts and a thinner filament between them. (Students may or may not notice the shunt.) They should also have noted that it was the filament that glowed. The blinker bulb drawings should include an additional metal piece between the two posts. The filament is connected to this piece and one of the metal posts. Note that this center piece is touching the other metal post, making a circuit through the filament. If students have difficulty, ask them what they notice about the color of the middle post. This center post is actually the bimetallic strip, and students may notice that it is brass-colored on one side (the side away from the filament) and silver-colored (stainless steel) on the other.

Explain

Safety note: Students should wear safety goggles while observing this demonstration, and the teacher should follow all safety rules for working with heat sources.

Have students share with the class their findings after having compared the two bulbs. In the simple bulb, the current flows in one of the posts, through

FIGURE 10.2	Bulb circuit setup

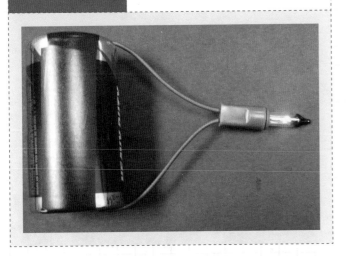

FIGURE 10.3	Close up (30×) of bimetallic strip in blinker bulb while off and on

the filament, and out the other post (Figure 10.1). In the blinker bulb, the current flows up the post to the bimetallic strip and then through the filament out the other post.

Use a bimetallic-strip demonstrator (available from vendors for about $6) to show students how it bends when heated. You can either use a flame or an incandescent 100W lightbulb with a reflective shade to heat the strip. It can also be heated on the top of a hot plate. Students will note that the strip bends away from the brass and toward the stainless steel. Now have students try to make predictions about the blinker bulb. You can refer students to Figure 10.3, which is a 30× magnification of the point where the post touches the bimetallic strip. Note in the photo that the bimetallic strip (on the right) is making contact with the post and that the bulb is lit. This causes the bulb to heat up and the strip bends slightly to the right, breaking the connection to the post as shown in the photo on the left. When this happens, the current flow is interrupted and the bulb goes out, which cools the strip enough for it to straighten and again make contact, repeating the process. We have included these photos because it is not possible to view this process with a hand lens.

Therefore, students will build a large model of this system in the Extend stage of this lesson.

Extend

In this stage, students will build a model of how the bimetallic strip works as a heat-sensitive switch in the blinker bulb. While there are multiple ways to construct such a model, Figure 10.4 on page 76 shows how we constructed one with basic lab materials. For each group of students, you will need the materials listed in Activity Worksheet 10.1 (p. 78). If students have difficulty, you may wish to set up a model of the switch.

It is important for students to realize that the bimetallic strip acts as a conductor in the bulb and that current flows through it. A simple way to start is for students to connect one end of the holiday light segment to one terminal of the D-cell battery. The other end of the bulb is connected to the bimetallic strip with an alligator clip. The second alligator clip connects the bimetallic strip to the D-cell battery, completing the circuit. The two clips on the bimetallic strip need to be relatively close to one another to compensate for the resistance of the strip. To include the nail, remove the second alligator clip from the bimetallic strip and

FIGURE 10.4 Model blinker bulb setup

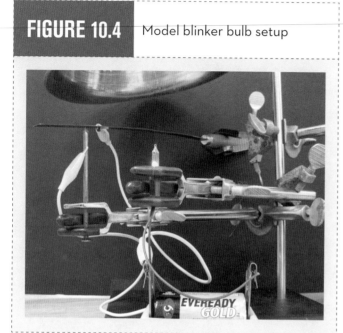

FIGURE 10.5 Close-up of closed and open bimetallic switch model

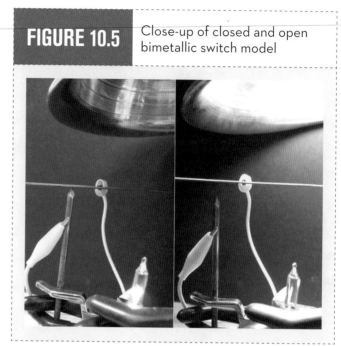

attach it to the nail. The light should go on each time a student touches the nail to the strip and off when the nail and the strip are separated. Finally, students need to clamp the bimetallic strip and the nail in such a way that when the lamp is turned on, it heats the strip so that it moves away from the nail, opening the circuit (Figure 10.5). Note that the side with the copper or brass will have to be touching the nail for this to work. Students should be able to trace the path of the current through the model heat-sensitive switch that is in the blinker bulb. The current flows from the cell to the nail that is touching the strip. The current flows through the strip to a wire, to one post of the bulb, through the filament to the other post, and returns to the other terminal of the cell.

Evaluate

Have students use the bulbs from the Explore stage to construct a circuit where the blinker causes the regular bulb to go on and off, as well. In order to do this, students will need to determine that the two bulbs need to be connected in series with one another. You may want to revisit the initial Engage demonstration of the entire light set blinking on and off. The current needs to flow from one bulb to the other. Therefore, when the blinker is off, there is no current flowing through the regular bulb, either.

Conclusion

The use of the bimetallic strip to create a heat-sensitive switch to make a bulb blink on and off is an example of an innovation of an invention. The invention of holiday lights has been modified through innovation many times over the years. We have focused on the engineering involved in making a holiday light blink on and off. Work in this area still continues as newer blinking holiday lights are operated by means of electronic controllers that can be programmed to make the lights blink in a variety of patterns. Engineering of holiday lights will likely continue to evolve in ways that Edison and Johnson would scarcely recognize.

References

Concannon, J. P., P. L. Brown, and E. M. Pareja. 2007. Making the connection: Addressing students' misconceptions of circuits. *Science Scope* 31 (3): 10–14.

Hoffman, J., and J. Stong. 2002. Electric connections. *Science and Children* 40 (3): 22–26.

International Technology Education Association (ITEA). 2002. *Standards for technological literacy: Content for the study of technology.* 2nd ed. Reston, VA: ITEA.

Moyer, R., J. Hackett, and S. Everett. 2007. *Teaching science as investigations: Modeling inquiry through learning cycle lessons.* Upper Saddle River, NJ: Pearson/Merrill/Prentice Hall.

Nelson, G. 2008. The antique Christmas museum. *www.oldchristmaslights.com.*

Sapp, L. 1999. Teaching circuits is child's play. *Science Scope* 22 (6): 64–66.

Series and Parallel Circuits

This lesson requires that students have some understanding of simple parallel and series circuits. In a parallel circuit, each resistance, or bulb in this case, has its own separate path for the flow of current. However, in a series circuit, the current flows through each resistance (bulb) one after another. If one bulb in a series circuit is removed, the remainder of the bulbs will go out because the pathway for the flow of current has been interrupted. This is not the case, of course, with a parallel circuit. In addition, early holiday series light sets would all go out if one of the bulb filaments burned out. This required one to search the entire string for the one burned-out bulb. Today's miniature series bulbs have a conductor (called a shunt) parallel to the bulb's filament that allows the current to continue to flow to the other bulbs in the event one burns out (see the wire wrapped around the two posts in the bulb on the left in Figure 10.1). For additional resources for teaching about circuits see Hoffman and Stong (2002); Concannon, Brown, and Pareja (2007); and Sapp (1999).

CHAPTER 10

ACTIVITY WORKSHEET 10.1   Investigating Blinking Bulbs

Safety note: Review safety rules with your students for working with batteries, bulbs, and wires before starting this activity.

Engage

1. Observe your teacher light the string of bulbs. What happens when one of the bulbs is changed?
2. Discuss with your group your ideas about what is happening.

Explore

In this activity, you will investigate the following question: How is a regular bulb similar to and different from a blinker bulb?

Safety note: Before starting, put on your safety goggles.

1. Using the short segments of a light string, set up one simple circuit with a D-cell battery for the regular bulb and another simple circuit for the blinker bulb.
2. Disconnect bulbs from the cells and observe each bulb carefully with a hand lens. How are they similar and different?
3. Draw a picture of the inside structures of each bulb.

Explain

1. Try to trace the flow of the current through each bulb from one end of the cell to the other. Add arrows to your drawing to show the path of the current.
2. Observe what happens to a bimetallic strip when your teacher heats it.

3. The center post in the blinker bulb is a bimetallic strip. Look at it closely and determine which way it will bend when heated.
4. What do you think will happen to the flow of current in the bulb if this post bends?

Extend

1. Obtain the following material for each group: one D-cell battery, a holiday light string segment with a regular bulb, a bimetallic strip, a nail, a ring stand, ring-stand clamps, a lamp with a reflector, and two wires with alligator clips. You will either need holders or electrical tape to connect the wires to the cells.
2. You are going to use these materials to make a heat-sensitive switch that will turn the bulb on and off in much the same way as the blinker bulb functions.
3. First, make a circuit that lights the bulb. The current must flow through the bimetallic strip.
4. Now try to design a circuit where the current also must go through the nail. Have the nail touch the bimetallic strip. Can you turn the bulb on and off by touching the nail to the bimetallic strip?
5. Finally, use the lamp to make the bimetallic strip bend in such a way that it makes contact with the nail when cool and no contact when it is heated.
6. Explain how your model of the bimetallic strip switch works.

Evaluate

Connect the two holiday light string segments together in such a way that both of the lights (the blinker and the regular) blink on and off. Draw the circuit and explain how it works.

CHAPTER 11

WINDMILLS ARE GOING AROUND AGAIN

DO ALL THINGS old really become new again? Depending on your age, your mental image of a windmill may be of the classic Dutch style, the ubiquitous American farm style of the 19th and early 20th centuries, or the giant-sized wind turbines often seen today grouped in massive farms. While the design has changed over the centuries, the basic idea has remained—some type of blade captures the energy of the wind in order to turn a shaft that does some kind of work, such as turning a millstone or turning a coil of wire in a magnetic field to generate electricity. Wind is reemerging as a clean and reliable source of energy—primarily for the production of electricity.

In this 5E Model lesson, students will construct a simple pinwheel-type windmill to test the power generated by different designs. Students will compare three- and four-bladed pinwheels made from manila folders or plastic report covers. This lesson addresses the ITEEA standard, "Energy is the ability to do work using many processes" (ITEA 2002, p. 162). In addition, the National Science Education Standards for the middle level include the standard "energy is transferred in many ways" (NRC 1996, p. 155). Energy from the Sun is transferred to wind energy, which becomes mechanical energy as the windmill turns, and then lifts a weight storing gravitational potential energy. Students will also observe a pinwheel connected

to a small electric motor (approximately $5), which generates electricity that will be used in turn to power a smaller motor (approximately $2; motors are available through most science supply distributors) to turn another pinwheel. Here, the mechanical energy of the windmill is converted to electrical by the first motor acting as a generator, and then back to mechanical by the second motor, which finally turns another pinwheel producing wind again.

Historical Information

People have used the energy of the wind for sailing purposes since as early as 5000 BC. Windmills were first used in China and the Middle East for pumping water and grinding grain around 200 BC. These ideas were later brought to Europe, where the Dutch improved on the design. The technology made its way to America and in the late 19th century was popular on farms and in rural areas, primarily to pump water and generate electricity. By the 1930s, most rural areas had been wired for electricity, and the use of windmills declined until the price of fossil fuels began to increase in the 1970s. Today, "wind energy is the world's fastest-growing energy source and will power industry, business and homes with clean, renewable energy for many years to come" (U.S. DOE 2005, p. 1).

FIGURE 11.1
Pattern for three- and four-bladed 225 cm² pinwheels

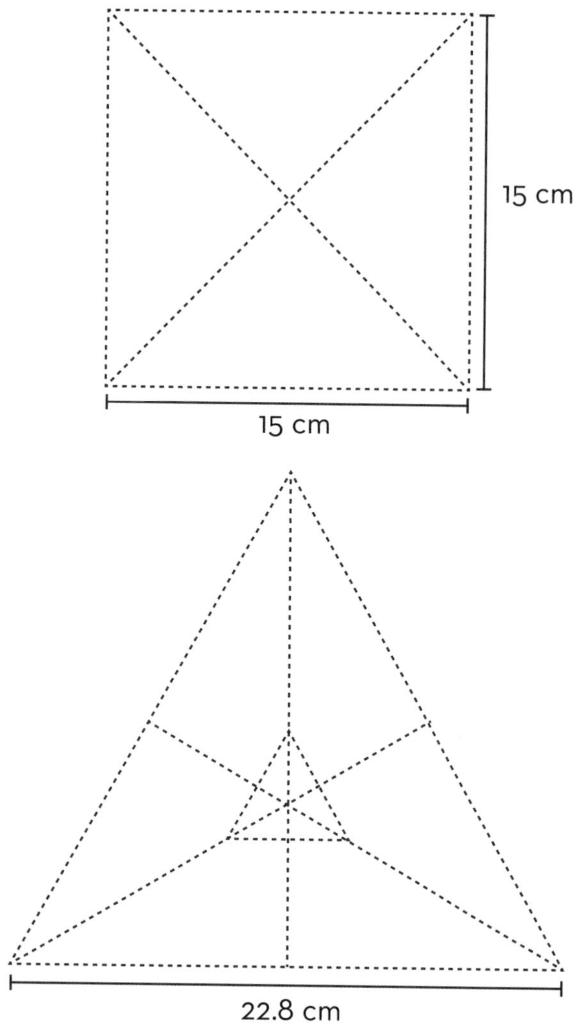

15 cm

15 cm

22.8 cm

FIGURE 11.2
Pinwheel details—the two longer straws are not glued to the pencil and act as bearings

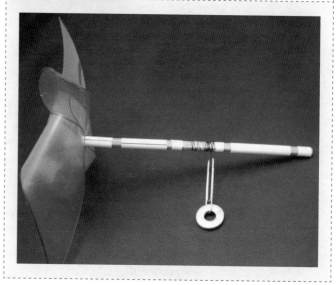

Investigating Windmills (Teacher Background Information)

Engage

Safety note: Review with students safety guidelines on the use of sharp objects such as scissors and pushpins. Students should use caution when working with glue guns and should wear safety goggles at all times during the investigation. Use low-temperature hot glue guns for this activity.

To engage students, you may want to show them a variety of windmill photos, easily found online by searching image search engines using the term *windmills*. Initiate a discussion of students' ideas regarding the function and purpose of windmills. Depending on your location, students may or may not have significant experience with windmills. You may also wish to have students construct a K-W-L chart with their ideas and questions about windmills.

Have students construct a pinwheel of their design. They should try to determine how moving air causes the blades to turn. This will provide some initial scaffolding experience in constructing a pinwheel before trying to make one that is capable of lifting a weight in the Explore stage.

Focus students' attention on the fact that windmills have differing designs and varying number of blades. Explain that in this lesson they will design an experiment to see whether a three-bladed or a four-bladed

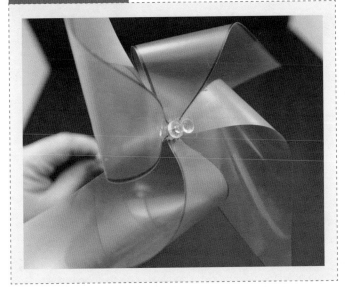

FIGURE 11.3 Completed pinwheel with detail of pushpin

FIGURE 11.4 Spinning pinwheel lifts the weights

pinwheel-type windmill can lift a weight faster. If you would prefer to have students participate in a much more open inquiry and design process, use the ideas in the Evaluate stage for the Explore component of the lesson.

Explore

Each group of two to four students will need two pencils with newer, flat-topped erasers, two large milkshake straws (8 mm rather than the standard 5 mm diameter), two file folders or plastic report covers, two pushpins, 100 cm of thread, two paper clips, four washers or other weights to hang on the paper clips, ruler, scissors, tape, stopwatch, template for the three- and four-bladed pinwheels (Figure 11.1), and access to a low-temperature hot-glue gun. If you want each group to find the mass of the weights and the paper clips, then they will need a balance, as well. One large box fan per group is ideal, although groups can share them. Students should be reminded to keep fingers and clothing away from the fan blade.

It will be helpful if you show students completed pinwheels (Figures 11.2 and 11.3). You may choose to give students the option to design their own pinwheels or use the procedure described in Activity Worksheet 11.1. Students may encounter some difficulty making the pinwheel turn a shaft in order to lift the weight. It is relatively easy to construct a pinwheel in which the blades turn; however, it takes much more tinkering and design work to construct a pinwheel-type windmill in which the blades freely turn and cause the shaft to turn, as well. Students must take care at all steps of the windmill construction. Wobbly shafts, bearings with too much friction, and unbalanced blades will all greatly interfere with the performance of the windmills. Your students will discover this whether you let them design their own windmills or follow the instructions provided in Activity Worksheet 11.1. Depending on your schedule, you may choose to have students construct and troubleshoot windmills during one class period and then test them the next. Small pieces can be stored in a zipper-type plastic baggie.

You may wish to have students design their own procedures for testing the three- and four-bladed

FIGURE 11.5   Sample data table for pinwheels

Type of pinwheel	Time (in seconds) to lift weight					
	Trial 1 (s)	Trial 2 (s)	Trial 3 (s)	Trial 4 (s)	Trial 5 (s)	Average (s)
Three blades	7.2	5.6	4.4	5	6.2	5.68
Four blades	1.8	2	1.8	3	2	2.12

pinwheels, or you can follow the sample procedure described here. Students should practice holding the pinwheel in front of a fan so that it spins freely and does not rub against their hand (Figure 11.4). Remind students to practice fan safety and to continue to wear their goggles. Students should hold the pencil level by the two straw bearings and not squeeze too tightly for maximum performance. For all trials, the pinwheel should be held the same distance from the fan and in the same relative position with respect to the fan. It is also advisable for students to practice using the stopwatch to time how long it takes to lift the weights. Students should conduct multiple trials and record their data into a table as shown in Figure 11.5.

Explain

While students' data will likely vary, they should all determine that the four-bladed pinwheels lift the weights faster. For a pinwheel of this size, the curved area facing the wind is greater for the one with four blades than three, since the three-bladed pinwheel has a greater surface that is flat. It is this curved area that produces the torque that causes the blades to turn.

Help students understand the energy transfer they have just experienced. The motor on the box fan has changed electrical energy into mechanical energy powering the fan, which moved the air and created wind. The wind was transferred to mechanical energy of the spinning pinwheel, which turned the pencil and lifted the washers. The washers now have stored gravitational potential energy. If the fan is turned off, gravity will pull down on the washers, and this potential energy is transferred to the kinetic mechanical energy of the turning pencil and pinwheel, and this mechanical energy is finally transferred by the pinwheel to the air again.

Have students continue to work with their groups for this portion of the lesson to facilitate learning for those who may have difficulty with the calculations. If your students are not familiar with the scientific definition of work, you may need to review. Work is the product of a force acting through a distance and is commonly measured in units of newton meters (Nm), also known as joules (J). It is likely that your balance will measure mass in grams rather than weight in newtons. Either you or your students will have to convert grams to newtons. Since $w = mg$, a force in newtons can be found by multiplying the mass in kilograms by the gravitational constant g which is 9.8 m/s². In our example, the mass of the paper clip and two washers was 10.4 g (0.0104 kg). Multiplying this by g results in a force of 0.102 N. We lifted the weights 40 cm, so the amount of work done is calculated as follows:

$$\text{work} = \text{force} \times \text{distance} = 0.102 \text{ N} \times 0.4 \text{ m} = 0.041 \text{ Nm}$$

Since students lifted all of the weights the same distance (and the weights were identical), both of the pinwheels did the same amount of work. However, the four-bladed pinwheel was able to do the work substantially more quickly (nearly three times as fast in our data) and therefore is more powerful. Power is a measure of the rate at which work is done. For the three-bladed pinwheel, students will find the following:

$$\text{power} = \text{work/time} =$$
$$0.041\text{Nm}/5.68\text{s} = 0.007 \text{ watts (W)}$$

For the four-bladed pinwheel, the power students will find is the following:

$$\text{power} = 0.041 \text{ Nm}/2.12 \text{ s} = 0.019 \text{ W}$$

FIGURE 11.6 Pinwheel with generator turning motor and another pinwheel

Extend

In this section, you will need to construct two pinwheels to attach to the two small motors using a short piece of a coffee stirrer and a dab of hot glue. The two motors are connected to each other with wires and alligator clips. When the pinwheel attached to the larger motor is placed in front of the fan, it will cause the second pinwheel to turn (Figure 11.6). To help students understand the transfer of energy, connect one motor to a 1.5-volt cell so that the pinwheel spins. The battery transfers stored chemical energy into electrical energy that the motor transfers into mechanical energy that turns the pinwheel. An electric generator is essentially a motor working backward—if you turn the shaft of a motor, this mechanical energy can be turned into electrical energy. This is what causes the second pinwheel to spin. To review the energy transfer: the wind causes the first pinwheel to turn, which turns the shaft of the larger motor, which produces an electric current that causes the second motor to turn as this electrical energy is transferred to mechanical, which turns the second pinwheel and causes the air to move, creating wind. Students may note that the second pinwheel is turning much more slowly due to inefficiencies of the transfers, with most of the "lost" energy going into heat from friction.

This lesson may be extended in many ways. Questions raised by students may deal with some of the current engineering, environmental, and societal issues related to the expanded use of windmills. Some engineering issues students could research are location of windmills relative to electrical needs and grid connections, initial cost compared to value of electricity generated, and long-term maintenance issues. Environmental questions might focus on potential harm to birds and bats and how the carbon footprint of the manufacture, transportation, installation, and maintenance of the windmill compares to traditional electrical power generation. Societal issues include what to some may be perceived as visual pollution and noise pollution, and their impacts on people and commerce.

Evaluate

There are many ways this investigation can be expanded, since there are many variables that can be tested by students: blade design (length, shape, and material), vertical versus horizontal shaft windmills,

or improving the soda-straw bearings to reduce friction loss. You may wish to have students search the internet and the library for information on windmill design (see KidWind Project 2010).

Conclusion

Wind energy is a reemerging source of renewable energy that has served humans for much of our history. Students may not realize that many engineers work on wind-related environmental projects, such as trying to maximize the amount of energy that can be transferred from the wind into electrical energy for use by society without the detrimental effects of burning fossil fuels.

References

International Technology Education Association (ITEA). 2002. *Standards for technological literacy: Content for the study of technology.* 2nd ed. Reston, VA: ITEA.

Kidwind Project. 2010. Wind turbine blade design. *http://learn.kidwind.org/learn/wind_turbine_variables_bladedesign*

National Research Council (NRC). 1996. *National science education standards.* Washington, DC: National Academies Press.

U.S. Department of Energy (DOE). 2005. History of wind energy. *www1.eere.energy.gov/windandhydro/wind_history.html*

Resources

Moyer, R. H., J. K. Hackett, and S. A. Everett. 2007. *Teaching science as investigations: Modeling inquiry through learning cycle lessons.* Upper Saddle River, NJ: Pearson/Merrill/Prentice Hall.

U.S. Department of Energy. 2005. Wind. *www.eere.energy.gov/topics/wind.html*

ACTIVITY WORKSHEET 11.1   Investigating Windmills

Safety note: Wear safety glasses during this activity. Use caution when working with the low-temperature hot glue gun, cutting, using sharp objects, and using the fan for testing.

Engage

1. Have you ever seen a windmill? How do you think a windmill works? For what purpose do you think people use windmills? What questions do you have about windmills? Record your thinking. Discuss your ideas with your classmates.

2. Your teacher will provide you with materials to use to design a pinwheel. Construct your pinwheel and try to figure out how moving air causes the blades to turn.

3. Now that you have some experience in designing and constructing a pinwheel, you will design a test to determine if the number of blades on a pinwheel-type windmill will affect its power. You will investigate the following question: Will the number of blades on a pinwheel affect how fast it can lift a weight?

Explore

1. Using the provided materials, construct the two pinwheels, one with three blades and the other with four. [Note to teachers: Alternatively, you could provide students with the following procedure and the pattern provided in Figure 11.1 with the same surface area of 225 cm² for each. Procedure: Cut out each pinwheel along the dotted lines. Note that each is cut in from the corners about two-thirds of the way to the center. Use the pushpin to make the holes at the dots in the corners and in the center. Insert the pin through the corners and then the center to secure, and set aside. Cut six pieces of straw as shown in Figure 11.2. Slide them onto the pencil and glue all but the two longer pieces, which serve

as bearings in which the pencil can freely spin. Use glue sparingly, as excessive glue will greatly impede the spinning of the pinwheel. Attach a 40cm length of thread with a small piece of tape to the center of the pencil as shown. A paper clip and two washers are attached to the other end of the thread. Finally, push the pin and pinwheel into the eraser, securing with a small dab of glue.]

2. Which pinwheel do you think will lift the weight faster, the three- or four-bladed pinwheel, if you hold each in front of a large box fan?

3. Plan a way to test the two pinwheels. In your plan be sure to include multiple trials to find an average time. Think also about which variables you must keep constant for a fair test. Share your plan with your teacher before you begin. Keep fingers and clothing away from the moving fan blades and continue to wear your goggles.

4. Time how long it takes each pinwheel to lift the weight and then organize your data into a table.

Explain

1. Share your results and conclusion with the class. How do your results compare with other group's? Did one pinwheel perform better than the other? What is your evidence for this? Look carefully at the two pinwheels and see if you can deduce why one lifted the weight faster.

2. How was energy transferred in the pinwheels you made? Where is the energy when the weight has been lifted all the way up to the pencil shaft? (Hint: If you turn off the fan, what happens to the weight and your pinwheel if you hold it gently by the straws?)

(Activity Worksheet 11.1 continues)

Activity Worksheet 11.1 continued

3. Work is the product of a force acting over a distance. If the force is measured in newtons (N) and distance in meters (m), then the amount of work is in newton meters (also called joules). The formula for work is the following:

$$w = f \times d$$

How much work did each pinwheel do? Your teacher will either give you the force in newtons or show you how to calculate it. The distance is the height your pinwheel lifted the weight.

4. If one of your pinwheels is able to do this work faster than the other, then it is said to be more powerful. Power (in watts) can be calculated by dividing the amount of work done, in joules (J), by the amount of time it took to do the work, in seconds:

$$power = work \div time$$

Calculate the amount of power in watts (joules per second) for both of your pinwheels.

5. Is one of your pinwheels more powerful than the other? How do you know? Compare your results with those of other groups in your class.

Extend

1. Observe what happens when the small motor attached to the pinwheel is connected to a battery. Explain the transfer of energy.

2. Predict what will happen if the motor is connected instead to a second, larger motor that is spinning in the wind from the box fan.

3. In the second case the wind caused the larger motor to turn. How did the system of the two motors connected together transfer this wind energy?

Evaluate

In this investigation, you found out what happened when you compared three- and four-bladed pinwheels. There are many other possible variables and windmill designs. Select a factor and design a test. Compare your results of the new design to the pinwheel model you investigated here. Determine the amount of power your design produces and share your work with your classmates.

CHAPTER 12

A LITTLE (FLASH) OF LIGHT

THE FLASHLIGHT IS often unappreciated until the electricity goes out. We all reach for one and hope that the batteries are still good. The flashlight is a simple device that is composed of a lightbulb, usually two cells connected in series, a housing, a switch, and a reflector for the light. All flashlights essentially use these parts to complete a circuit that converts the stored chemical energy in the cells to light energy.

In this lesson, students will take apart an inexpensive flashlight (less than $2 with D-cell batteries) to determine how the parts work together to produce light. Students will then make a flashlight of their own design using common household materials. While constructing a simple circuit with a switch is straightforward, putting it all together as a device is likely to challenge students. Many middle level students may have little experience designing and building something on their own. The ITEEA recommends that students understand the core concept of dealing with malfunction: "Malfunctions of any part of a system may affect the function and quality of a system"(2002, p. 39). In the case of a flashlight, there are many possible reasons why it may fail (malfunction)—a burned out bulb, a dead cell, or a poor connection, to name a few. Resolving problems while building something may be difficult for some students; consequently, you may need to encourage them to be analytical in their attempts to problem solve. Like all lessons in the book, this one follows the 5E Model.

The science content of this lesson focuses on simple circuits and the transfer of energy from one form to another: "Electrical circuits provide a means of transferring electrical energy when heat, light, sound, and chemical changes are produced" (NRC 1996, p. 155). It is recommended that students have a basic understanding of simple electrical circuits before attempting this lesson. If you need ideas for a review of switches and simple circuits, please see Hoffman and Stong (2002); Concannon, Brown, and Pareja (2007); and Sapp (1999).

Historical Information

The flashlight epitomizes the engineering principle of taking advantage of recent discoveries to invent new devices that people can use to solve problems. The essential parts of the flashlight—dry cell batteries and lightbulbs—were both invented in the late 1800s. The lightbulb was invented by Thomas Edison in 1879, and the dry cell was invented by Carl Gassner in 1888 (Schneider 1996). Prior to the invention of the flashlight, people used lanterns or candles for portable

FIGURE 12.1 Inside of flashlight

illumination. The first commercial flashlight became available in 1896 and was marketed by Conrad Hubert, who had a business selling electrical novelty items such as electric tie tacks and stickpins. The cells were weak and the early bulbs that used carbon fiber filaments were dim, so the lights could only be flashed on for a short time, which gave rise to the term *flashlight*.

Investigating Flashlights (Teacher Background Information)

Engage

Safety note: Review safety rules with your students for working with batteries, bulbs, and wires before starting this activity. Safety goggles should be worn for the entire activity.

For the Engage activity, each group of two to four students will need an inexpensive flashlight (check your local dollar store for models that use two D-cell batteries). Do not use rechargeable cells because they produce more current and can become dangerously warm if short-circuited. When batteries die, they should be recycled according to local hazardous-waste protocol, not thrown in the trash. In most states, common carbon-zinc cells are not considered hazardous waste.

Prior to the activity, initiate a discussion by showing students a flashlight and inquiring about their previous experiences and prior knowledge. Focus the discussion on how a flashlight works and the circuitry involved. Have the groups disassemble and diagram their flashlights. Usually, inexpensive flashlights have a metallic strip that completes, makes, and breaks the circuit to turn the lamp on and off (Figure 12.1). Allow 10–15 minutes for this initial investigation. Have students put flashlights back together so that the flashlights may be used again with subsequent classes.

Explore

Allow most of a class period for constructing and comparing student flashlights. Provide the following materials for each group or pair of students to make a circuit similar to the one found in a flashlight: two D-cell batteries, a flashlight bulb, electrical tape, and two pieces of bell wire with the ends stripped. A switch can be fashioned out of two paper fasteners, a piece of cardboard, and a paper clip (Figure 12.2). Be sure all students understand this simple circuit before proceeding to the construction of a flashlight. At this point, provide a variety of materials for students to use in making their flashlights. Paper towel (or toilet paper) tubes can be used for the body. Because students may wish to keep their flashlights, instruct them to bring tubes prior to the activity. Bulb holders can be made from small plastic cups (89 ml or 3 oz.) or the top portion of half-liter water bottles. Aluminum foil can be used to make reflectors. Additional typical schoolroom supplies, such as glue, rubber bands, and the like should be provided. Provide a tray of supplies for each group of two to four students. Some groups may ask for additional wire depending on their design. Sample student flashlights are shown in Figure 12.3.

Again, have students make a sketch of their design that explains how the circuitry functions. Students can

FIGURE 12.2 Circuit and switch

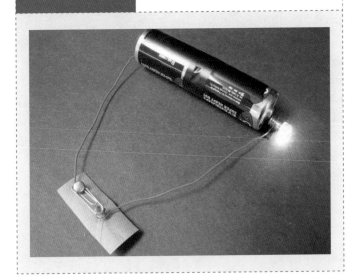

FIGURE 12.3 Sample student flashlights

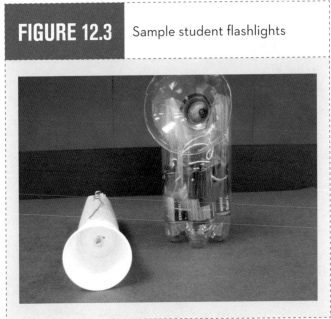

add arrows to their sketch to show the flow of energy through the flashlight. The dry cells contain stored chemical energy that changes to kinetic energy in the form of electric current, which is transformed into light and thermal energy as it passes through the high resistance of the filament in the lightbulb.

Explain

Have students compare their sketches of the manufactured flashlight with their own designs. The circuits will likely be very similar, unless an unusually creative design was employed. However, they may also differ in a number of ways. For example, most commercial flashlights use a metallic strip and a spring, rather than wires, to complete the inner circuitry. Also, the switch will usually slide and the actual connections are inside the flashlight. Depending on the creativity of the student designs, there may be additional differences, as well.

Students are likely to encounter a number of problems (malfunctions) as they construct the flashlights. Common problems include holding the bulb in place, loosening wire connections, mating the tip of the bulb to

the end of the cell, and improper aligning and ordering of the cells. Have students share their flashlights with the class and explain and demonstrate how they work. Such sharing helps to create a learning community that enables students not only to share with one another but also to see creative solutions to problems that they may not have considered themselves. Indeed, this is the same process that engineers (and scientists) experience when they deliver papers at conferences and publish their work in journals.

Extend

Gather an assortment of different types of flashlights—Maglites, lantern battery-style lights, penlights, small LED flashlights, etc. (See Figure 12.4 on page 90.) Have each group examine one of the flashlights to determine how it works. These flashlights can be reused for each of your classes. Some flashlights may have a push-button switch, some may have a twist switch, and others will have a sliding switch like the ones used in the activity. The flashlights will also use different types of batteries—the lantern style uses a 6-volt battery, the penlights may use AA- or AAA-cell batteries, and some

CHAPTER 12

FIGURE 12.4   Different types of flashlights

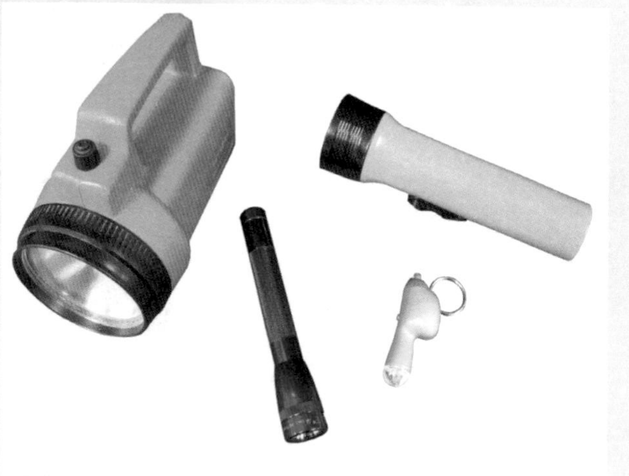

may use C-cell batteries. Small LED lights often use button cells. Once again, have students draw a sketch to show the circuitry in their flashlight. Students can also draw their sketches on a transparency so that they can discuss each flashlight type with the entire class. You may also wish to have students conduct some research on the history of flashlights (see Schneider 1996). In addition, students could research flashlights that are used for special purposes—red LED flashlights are used for night vision, as well as for military purposes.

To further extend the lesson, you may want to have students compare a two-cell LED flashlight with a two-cell regular penlight to see if the LED light uses less current. Put fresh cells in each and leave them on until the cells are depleted. You could also calculate the ratio of cost to time for each so that students can see that, eventually, the more expensive LED light is actually cheaper.

Evaluate

Show students a 9-volt battery and ask them to design and draw a sketch of a flashlight that uses a 9-volt battery. Have students label the main parts of their flashlight plan and write a short explanation of how it works. The drawings should show the circuitry, including a switch that will enable the bulb to be successfully turned on and off. If students have not had much experience with designing, it is not recommended to base a grade on the creativity of the design. As students gain experience and confidence in trying out their own ideas, you will begin to see a greater variation in student designs. Note: It is not recommended that you have students actually build 9-volt flashlights because you will need higher-voltage bulbs. Due to the proximity of the terminals, it is more likely for students to create a short circuit that may become quite warm.

Conclusion

Engineering is often about the simple devices that we take for granted in our everyday lives. The invention of flashlights allowed people to ride bicycles at night, search in dark closets, and move around safely during power outages (Schneider 1996). While there have been improvements in cells, bulbs, and switches, the design of the basic flashlight has changed little in over 100 years, especially when compared to changes in other technologies—air travel from the Wright brothers to the space shuttle, for example.

References

Concannon, J.P., P.L. Brown, and E.M. Pareja. 2007. Making the connection: Addressing students' misconceptions of circuits. *Science Scope* 31 (3): 10–14.

Hoffman, J., and J. Stong. 2002. Electric connections. *Science and Children* 40 (3): 22–26.

International Technology Education Association (ITEA). 2002. *Standards for technological literacy: Content for the study of technology.* 2nd ed. Reston, VA: ITEA.

Moyer, R., J. Hackett, and S. Everett. 2007. *Teaching science as investigations: Modeling inquiry through learning cycle lessons.* Upper Saddle River, NJ: Pearson/Merrill/Prentice Hall.

National Research Council (NRC). 1996. *National science education standards.* Washington, DC: National Academies Press.

Sapp, L. 1999. Teaching circuits is child's play. *Science Scope* 22 (6): 64–66.

Schneider, S. 1996. Flashlight museum. *www.wordcraft. net/flashlight.html.*

ACTIVITY WORKSHEET 12.1   Investigating Flashlights

Safety note: Students should wear safety glasses during the activity.

Engage

1. Carefully take apart a flashlight to determine how it works. Caution: Do not attempt to take the cells or the lightbulb apart.

2. How are the parts connected in order to make a complete circuit?

3. Make a sketch of your findings that shows how the bulb lights. Be sure to include how it is turned on and off.

4. Now that you have examined a manufactured flashlight, how would you design your own?

Explore

1. Construct a circuit with two cells in a series that will illuminate a lightbulb.

2. Add a paper-clip switch to your circuit to turn the bulb on and off.

3. Using the materials provided by your teacher, design and then construct a flashlight. Your flashlight should include a switch. Exercise caution with sharp objects and projectiles.

4. Make a sketch of how your flashlight works. Show how it is turned on and off, as well.

5. Trace the path of the energy flow in your flashlight, starting with the stored energy in the cells.

Explain

1. Compare the two sketches of the manufactured flashlight and the one of your own design. How are they the same? Different?

2. What problems did you encounter trying to build your flashlight? How did you solve these problems?

3. Share your flashlight with the class. Demonstrate how it works. What problems did other students encounter and how did they solve these problems?

Extend

1. Look at the different flashlights that your teacher has provided.

2. Take one flashlight apart to determine how it works, including how it is turned on and off.

3. Draw a sketch that shows the circuitry in this flashlight and share it with your class.

Evaluate

Your teacher will show you a 9-volt battery. On paper, design a flashlight using this battery. Show and include a switch in your plan. Label all of the parts.

PART 6

Outdoor Recreational Engineering

CHAPTER 13

LIFE PRESERVERS—INCREASE YOUR *v* TO LOWER YOUR *D*

THE LAST TIME you were on a boat or at the beach, what sort of life jackets—known as *personal floatation devices* (PFDs) to the U.S. Coast Guard—do you recall seeing? At the beach, you may have seen children wearing inflatable water wings (not an approved Coast Guard device) on their upper arms. Some children's bathing suits have floatation devices built in. On a boat you may recall seeing bulky, bright-orange jackets or sleeker ski vests. By regulation, recreational boats must also carry a throwable floatation device of some kind, usually a ring or a seat cushion. It is clear that all of these things have been designed to help keep a person afloat. Have you ever wondered exactly how PFDs work?

In this 5E Model lesson, we consider how life jackets are designed and the physics of how they work. Different life jackets have different functions, and, therefore, engineers have had to design for these different uses. This lesson centers on a design challenge for students to plan and construct a model life jacket for a plastic action figure that will keep the figure floating faceup using the smallest amount of life jacket material. One of the ITEEA standards for middle-level students states, "Requirements for a design are made up of criteria and constraints" (ITEA 2007, p. 95). This

same idea is also included as a middle-level grade band endpoint in *A Framework for K–12 Science Education* that includes engineering as a core idea: "The more precisely a design task's criteria and constraints can be defined, the more likely it is that the designed solution will be successful" (NRC 2012, p. 205). In our design challenge, the criterion is to keep the figure floating with its face out of the water, and the constraint is to use the least amount of material.

In addition, students will learn how the overall density of the action figure changes with and without the life jacket, and ultimately realize that a life jacket essentially increases the wearer's volume while adding insignificant mass, thus resulting in reduced density.

Density is a difficult property for many students to comprehend for two reasons. First, density, like boiling point and solubility, is an extrinsic property; that is, it is independent of the quantity of material being considered (NRC 1996). Second, density (mass per unit volume) is a ratio that increases with increased mass but decreases with increased volume. Many students have difficulty conceptualizing the inverse part of this relationship—that increasing volume decreases density. For these reasons, density should be visited numerous times during the middle-level years.

Historical Information

The case for wearing a personal floatation device when engaged in water recreation could not be stronger. According to the U.S. Coast Guard, most drownings occur in inland waters and often close to help. The overwhelming majority of people who drown each year were not wearing a life jacket of any kind (USCG 2008). The key to water safety is simply wearing a life jacket. Although many people do not wear life jackets because they feel the jackets are bulky or cumbersome, there are a number of different Coast Guard–approved types of life jackets or vests—for offshore use, for near-shore use, and for special purposes. Another type is designed to be throwable to someone in the water. Since the more-comfortable vests were introduced around 1970, the number of drownings in the United States has dropped from about 1,500 to 500 per year (USCG 2008).

Prior to modern life jackets, people made use of a variety of devices to assist in staying afloat if cast overboard. The earliest floatation devices were likely inflated animal bladders or chunks of wood. One of the earliest recorded life jackets was made of cork attached to canvas and invented by a Captain Ward in the United Kingdom in 1854 (RNLI 2011) (see Figure 13.1). Later, a vegetable fiber known as *kapok* was used to make life jackets. Kapok, which does not absorb much water, made an excellent buoyant material for life jackets and was used until synthetic fibers began to dominate in the 1960s.

Engineers have continued to tweak life jacket design, and there are now numerous variations. A 2011 contest, the Innovation in Life Jacket Design Competition, had the goal of trying to improve on the basic life jacket. The winning entry was a T-shirt with a built-in inflatable bladder that greatly increases the comfort of wearing a floatation device and thus the likelihood that one does (PFDMA 2011). Life jacket design is still evolving to reduce bulk and thus encourage people to wear the life jackets and to continue to vigorously partake in recreational activities.

FIGURE 13.1   The first cork life jacket

Investigating Life Jackets (Teacher Background Information)

Materials

Each student will need two small (about 4 or 5 cm tall) plastic action figures that are available at dollar stores in bags of 20 for $1. We used small soldiers and removed the weapons and the bases with scissors. Most action figures will have a density of just over 1 g/cm^3 and thus just barely sink. You should test to assure that your figures do indeed sink. Each student will also need a small piece of flexible (not polystyrene) foam material. The exact size does not matter, but about 7 cm × 7 cm

× 1 cm will work. We found that swimming noodles worked very well, as did the packing material used to ship new computers or electronic equipment. This material can easily be cut with scissors to any size or shape. The action figures and the foam will need to be replenished for each new class.

Each group of three to five students needs one container of water, scissors, a balance that measures to the nearest hundredth of a gram, and a graduated cylinder of sufficient width that the action figure will fit (and measures at least to the nearest milliliter). If necessary, you may need to trim the action figures in order for them to fit in the cylinder. You could have one low-temperature hot glue gun per group or choose to have just one gluing station that you monitor. For the Engage section, you will need two navel oranges, one peeled and one not. They can be reused for subsequent classes.

Engage

Safety note: Students should wear safety glasses during this activity and use caution when working with the low-temperature hot glue gun.

To get students thinking about sinking and floating, we suggest a simple demonstration. Like many fruits, a navel orange will float in water. However, it may surprise many of your students that when peeled, the orange will sink while the peel floats (see Figure 13.2). In this lesson, students will discover that a life jacket aids in floating in essentially the same way that the peel causes the orange to float—by increasing the volume of the orange without substantially increasing its mass.

The engineering challenge in this lesson is for students to design a way to make the action figure float with its face out of the water, regardless of how it is dropped into the water. That is, if it goes in facedown, the life jacket will cause the action figure to flip over to a faceup position. During the brainstorming session, encourage students to consider different life jacket designs with which they might be familiar. Stress to

FIGURE 13.2 Peeled and unpeeled navel oranges

them, however, that more divergent thinking often pays off with innovative designs.

Explore

The life preservers for the action figures can be fashioned out of the foam material, which students can safely cut with ordinary scissors. Depending on the number of glue guns you have available and the amount of supervision needed, you can either distribute one per group or set up one classroom gluing station. One station is sufficient, as there is little gluing required for this challenge. Because of the open-ended nature of this activity, you should preapprove written designs before students begin construction. Also, if digital cameras are available, you might want students to take photos of their designs in the water in order to assist in reporting results to classmates. You may wish to limit the number of times students redesign their life jacket, because they will be making another in the Extend stage of the lesson.

Explain

Have students compare successful and unsuccessful designs (see Figure 13.3, p. 98). They should conclude that in order for the action figure to float faceup, there must be more buoyant material on the upper portion of the body and also more on the front side. The former

FIGURE 13.3 Floating action figures with different life preserver designs

FIGURE 13.4 Measuring action figure volume

will assure that the head is upright and the latter that it is faceup.

At this point, you may need to review the basic concept of density. Students may be able to state that objects that are less dense than water (< 1.0 g/ml) float, but likely will find it more difficult to explain why. Thus, students should calculate the density of an action figure with and without a life jacket. It is critical that they carefully measure the volume using a graduated cylinder with units no larger than 1 ml (see Figure 13.4). Typically, the plastic action figures will have a density very close to 1; ours were about 1.05 g/ml.

From their data table (see Activity Worksheet 13.1, p. 101), students should be able to observe that the mass of the action figure with the life jacket changed very little; ours increased from 3 g to 3.15 g. Likewise, they should easily see that the volume increased a good deal; ours increased from 2.9 ml to 4.1 ml. Help students conceptualize that this increase in volume is critical to how life jackets work—they increase volume

substantially with very little increase in mass. Show students the formula for density $D = m/v$ and discuss what happens to the value of the fraction if v increases. Inverse relationships, such as the one between density and volume, are universally difficult for many students.

Extend

Most engineering problems, including the design of a life jacket, present engineers with constraints. In this case, a major constraint is the amount of material used to construct the life jacket. This is due in part to material cost, but also to try to reduce the bulkiness of the jacket, because that plays a significant role in whether people actually wear them. Thus, now that students understand the principles involved in keeping the action figure faceup in the water, they need to alter their design to use the least amount of material. Again, have students calculate the density of the action figure with the new design (see Figure 13.5). In this case, the density of their design will likely be higher

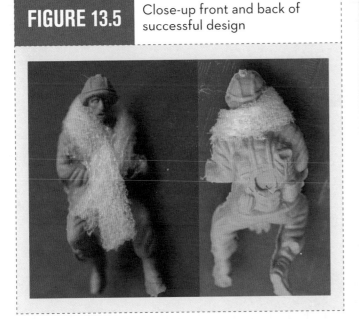

FIGURE 13.5 Close-up front and back of successful design

FIGURE 13.6 Infant life jacket design

than their first design and therefore closer to 1 g/ml. This is because using less foam material will cause the volume to be less and therefore the denominator of the density fraction will also be smaller, resulting in a larger value for *D*.

Evaluate

Show students a picture of an infant life jacket (see Figure 13.6) and ask them to analyze the criteria and constraints the engineer had to consider in its design. The major design criterion is that the jacket needs to be virtually fail-safe in its operation because babies are essentially helpless in the water. Given the constraint that babies often have weak neck muscles results in a design criterion that the jacket has to assure the baby's head is held up and out of the water. This complicates the design because extra material in the back of the life jacket will tend to cause the baby to float facedown. This must be compensated for with additional buoyant material on the front of the jacket. The flap behind the

head helps deal with another constraint, as well—babies are a bit top heavy due to the size of their head relative to their body, as compared to an adult. Thus, the flap at the top helps assure that the baby will float head up.

Conclusion

A Framework for K–12 Science Education (NRC 2012) includes engineering as both a content idea and a practice. This lesson integrates all of the STEM disciplines, while focusing on the core ideas of criteria and constraints in engineering and the practice of engineering design. The concept of density combines both science and mathematics. The product of engineering is technology. In this case, the technology developed was a device to keep a person faceup and afloat.

So, because you float better with a life jacket, remember to wear one the next time you are in a boat and note that even something as common as a life jacket was indeed engineered.

References

International Technology Education Association (ITEA). 2007. *Standards for technological literacy: Content for the study of technology.* 3rd ed. Reston, VA: ITEA.

National Research Council (NRC). 1996. *National science education standards.* Washington, DC: National Academies Press.

National Research Council (NRC). 2012. *A framework for K–12 science education: Practices, crosscutting concepts, and core ideas.* Washington, DC: National Academies Press.

Personal Floatation Device Manufacturers Association (PFDMA). 2011. Life jacket designs break new ground. *PFDMA Industry News. www.pfdma.org/news/default.aspx.*

Royal National Lifeboat Institution (RNLI). 2011. *History of the RNLI. www.rnli.org.uk/assets/downloads/historyfactsheet.pdf.*

United States Coast Guard (USCG). 2008. PFD selection, use, wear, & care. *www.uscg.mil/hq/cg5/cg5214/pfdselection.asp.*

Resources

Moyer, R., J. Hackett, and S. Everett. 2007. *Teaching science as investigations: Modeling inquiry through learning cycle lessons.* Upper Saddle River, NJ: Pearson/Merrill/Prentice Hall.

ACTIVITY WORKSHEET 13.1 Investigating Life Jackets

Engage

Safety note: Students should wear safety glasses during this activity and use caution when working with the low-temperature hot glue gun.

1. Do you think a navel orange will sink or float? Discuss with your classmates and record your prediction. Observe what happens when your teacher puts the orange in a container of water. How does this compare with your prediction?

2. What do you think will happen to the orange if it is peeled before being placed in the water? What about the peel? Again, discuss with your classmates and record your predictions for each part of the orange. Observe what happens and compare to your predictions.

3. Predict what will happen if you place your action figure into the water, and then test your prediction. How does this compare to the orange?

4. In this activity, your challenge is to design a way to make the action figure float with its face out of the water. Discuss your ideas with your classmates.

Explore

1. Considering the materials provided by your teacher, make a plan for a life jacket for your action figure that will keep its face floating out of the water.

2. After your teacher approves your plan, you can begin constructing the life jacket. Use caution when handling the glue gun to attach the jacket to the action figure.

3. Test your design by placing the jacketed action figure in a container of water with its face down. Take a photo or make a sketch of the action figure in the water.

4. Compare the photos (or sketches) of those designs that met the challenge criterion with those that do not. How do they differ?

5. If necessary, modify your design and retest.

Explain

1. Share your designs and the results of your testing with your classmates.

2. Now that you have seen many designs, what can you conclude about why the successful designs work?

3. Based on what you know about density, what can you infer about the density of the action figure compared to the density of the water before you made a life jacket for it? After you made a life jacket for it?

4. Calculate the density of an action figure with and without its life jacket. Which changes more, the mass or the volume?

	Mass	Volume	Mass/volume
Without jacket			
With jacket			

5. How does the life jacket cause the density of the action figure to vary? Compare this to how the peel affects the density of a navel orange.

6. You may already know that the density of water is equal to 1 g/ml. How do the densities of your action figure compare to water's density?

7. What can you conclude about the density of things that float?

Extend

1. One constraint an engineer has when designing a life jacket is the amount of material the design uses. Brainstorm some ideas why you think this is the case.

2. With this constraint in mind, design a second life jacket that uses the least amount of material but still meets the initial criterion of keeping the action figure's face out of the water.

3. Calculate the density of your action figure with this second life jacket design.

4. Compare this density with the density of the action figure with your first design. Which one is closer to the density of water? Why?

Evaluate

Look at a picture of an infant life jacket. Compare this life jacket with your best design. Explain what criteria and constraints you think the engineer had to consider in the design of the infant's life jacket.

CHAPTER 14

AIN'T SHE SWEET— BATS, RACKETS, GOLF CLUBS, AND ALL

THE PITCHER THROWS the ball and the batter takes a mighty swing. Crack! The ball is hit on the sweet spot and soars to the outfield. With the player on base, the next batter swings at the pitch. Thud! This time, the ball dribbles along the infield ground, and the batter's hands sting. Everyone who has played baseball or softball has probably experienced both of these outcomes. This may not seem like science, but it is a lesson in frequencies of vibrations and transfer of energy: "For most engineered sports equipment, energy transfer is the single most important scientific concept" (University of Colorado 2006, p. 1). In the case of a bat, the objective is to transfer the maximum amount of energy to the ball. When the ball hits the sweet spot, the vibrations set up in the bat are minimized, and therefore more energy can be transferred to the ball. However, when the ball hits near the handle—or anywhere other than the sweet spot—the bat vibrates more, and less energy is transferred to the ball. Transfer of energy is a major content standard for the middle level in physical science in the *National Science Education Standards* (NRC 1996).

Sports engineers work with governing bodies on safety issues and the development of training equipment. Helmets and sports padding must be able to absorb energy and not transfer it to the player. Con-

versely, engineers design bats, rackets, and golf clubs so that they can transfer maximum energy to the ball.

The sporting industry has become big business and precipitated the need for engineers to design equipment, because athletes want to be able to run faster, hit farther, and catch better. The ITEEA standards for middle-level students note that "throughout history, new technologies have resulted from the demands, values, and interests of individuals, businesses, industries, and societies" (2002, p. 77). A prime example is how tennis shoes have changed in the last 50 years from canvas in either white or black to the multitude of options available today. In baseball, uniforms have changed from wool to synthetics; helmets and other safety equipment for catchers and umpires have been developed; and bats have evolved over time from wood to metal and, perhaps in the future, back to wood.

Historical Information

Baseball became popular in the United States in the 1850s. Early bats came in many shapes and sizes, including flat sticks, as the game was derived from the British game of cricket. In 1859, bats first became standardized, with the diameter limited to 6.35 cm (2.5 in.). In 1869, the length of bats was limited to

CHAPTER 14

FIGURE 14.1 Finding a bat's sweet spot

FIGURE 14.2 Exploration setup

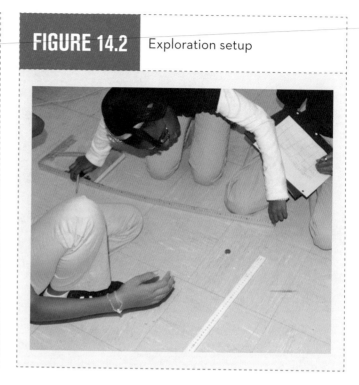

107 cm (42 in.), and in the 1890s, bats could no longer be flat on one side, and the maximum diameter was increased to 7 cm (2.75 in.). While metal bats were first patented in 1924, they did not become popular until the 1970s, when youth leagues considered them safer because they did not shatter easily. Because they are more durable, they are cheaper in the long run. However, advancements in the technology of metal bats have resulted in batters being able to hit the ball with greater force. Thus, the ball leaves the bat with greater speed, which can be dangerous to the pitcher, whose reaction time may not be quick enough to avoid serious injury. For this reason, some amateur baseball leagues are now considering changing back to wooden bats. Most wooden bats are made from ash trees, but some players are changing to maple after a player set a home run record with a maple bat. Professional baseball teams have always used wooden bats.

Investigating Sweet Spots (Teacher Background Information)

Engage

Each group will need a bat and a small hammer. You may be able to borrow this equipment from physical education and technical education teachers at your school. If you determine that there is any risk of flying splinters, have students wear safety goggles accordingly. Initiate a discussion about student experiences batting balls. Address clean hits as well as hits that shake the bat and sting one's hands. Some students may be familiar with the term "sweet spot" and may or may not understand its meaning. The term will be clarified as students work through the investigation. Students will likely report many variations between the two extremes of "crack" and "thud."

Explain to students that one of them will hold a bat by the handle end (Figure 14.1), while another student lightly taps it with a hammer starting at the bottom and

FIGURE 14.3 Sample results for the meterstick bat

Location on meter stick (cm)	Distance nickels move					
	Trial 1 (cm)	Trial 2 (cm)	Trial 3 (cm)	Trial 4 (cm)	Trial 5 (cm)	Mean (cm)
90	16	16	22	17	19	18
80	18	23	27	29	33	26
70	10	6	8	10	10	8.8
60	8	7	7	6	6	6.8

working upward. Have students predict what they will hear, feel, and see. Students should be asked to explain their answers based on their prior knowledge.

Students should report that they feel vibrations of varying degrees along the bat except for one spot—the sweet spot—that is about 15 cm (6 in.) from the barrel end of the bat (this may vary with different bat lengths). Students will also be able to see the bat vibrate and may note a different sound when the sweet spot is struck. If you do not have enough bats available for each group of four students, then conduct this introductory activity as a demonstration with one student.

Explore

Each group of four students will need two metersticks, two nickels taped together, and an object such as a textbook to support one end of a meterstick (Figure 14.2). The first meterstick is used as a model of a bat, and the second is used to measure how far the nickels travel. Ask students to predict if a ball will travel differently when it is struck at various points on the bat. Encourage them to construct a fair test to find out. One possible procedure would be to hold the meterstick tightly against a book, then place the nickels at the 90 cm mark at the other end of the meterstick. Pull the free end of the meterstick back 5 cm and release so it strikes the nickels. Measure and record how far

the nickels travel. Complete four more trials and find the average distance. Then repeat with the nickels positioned at the 80 cm, 70 cm, and 60 cm marks. A sample data table is shown in Figure 14.3.

Explain

Students should find that the distance the nickels travel increases as they move away from the end of the barrel to a point, and then decreases again as they move toward the handle of the meterstick. As noted in Figure 14.3, our nickels traveled the farthest when struck at the 80 cm mark that is very near the sweet spot of the meterstick (which we will determine in the next section to be at approximately the 75 cm mark). While this pattern should hold, the exact results may vary somewhat from group to group, depending on how they set up their procedures. You may wish to have students watch the short informative video Testing Bats (see Resources), which shows engineers and technicians working in a lab evaluating bats.

There are several different definitions of the *sweet spot*. The most common definition is the spot where the batter feels the least amount of vibration when the ball is struck, which is what we are determining when students tap with the hammer. Another definition is the spot where the ball will travel the farthest, which is very close to the spot of minimal vibration. Another

FIGURE 14.4 Finding the sweet spot of a meterstick

FIGURE 14.5 Finding the sweet spot of a golf club

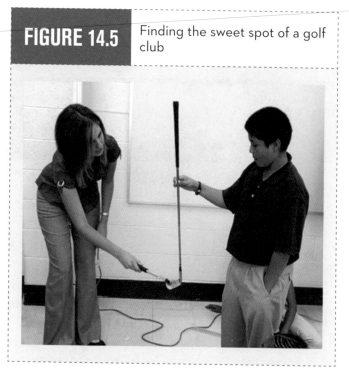

definition relates to the center of mass of the bat. All of these fall within the same area on the bat—the area known as the sweet spot.

Student data provide evidence that there is one spot on the meterstick bat that transfers the most energy to the nickels. At other locations, some of the energy is transferred to the increased vibrations of the meterstick bat, and thus the nickels do not travel as far. This is also what students experienced in the Engage activity when they felt the increased vibrations of the bat everywhere except the sweet spot. This is because when an object is struck, vibrations are set up in that object. At some points, called nodes, one of which is the sweet spot, these vibrations cancel each other out (Doherty 2007).

Because the ball causes fewer vibrations in the bat when it strikes the sweet spot, one hears a cleaner sound—the "crack" sound. When the ball hits the bat at other locations, it causes many more chaotic vibrations, and thus the "thud" sound is heard. It is akin to playing a chord on a piano as opposed to random notes.

At this time, you may wish to discuss the idea

that a major purpose of sports equipment design has to do with the transfer of energy. Clearly, bats are designed to maximize the energy transfer to the ball. Ask students to explain in terms of energy transfer the major design purpose of a helmet and other protective sports equipment.

Extend

Students should notice that the vibrations are more pronounced in the meterstick than in the bat, and they should have little difficulty locating the sweet spot—the area where those vibrations are minimized. For a standard wooden meterstick, this should be approximately 22 to 25 cm from the barrel end. Students should note that this corresponds closely to the location that resulted in the nickels going the farthest when struck by the meter stick.

Evaluate

To assess their understanding of sweet spot location, have students think about other sporting equipment, such as tennis rackets or golf clubs. You will need one piece of equipment for each group of students—a set of golf clubs will work nicely. Students should devise a plan to locate the sweet spot using a hammer in much the same way as they did with the bat and the meterstick (Figure 14.4). With a golf club, the sweet spot is located at the center of the clubface. Students should be able to feel and possibly see the vibrations when tapping on the clubface. If they tap to the left or right of center, they will also notice a twisting motion of the club shaft. In addition, they should notice a distinctive change in pitch when striking the club's sweet spot (Figure 14.5).

It is a bit more difficult to determine the sweet spot of a tennis racket with a hammer. One reason for this is that on modern rackets, the sweet spot is rather large and extends in an oval shape from near the bottom to above the center. Again, students should notice vibrations when the racket is tapped near the edges. Like the golf club, the racket will twist when struck to the left or right of center. You might also have students drop a tennis ball onto the racket at various points and observe the ball's bounce.

As you discuss the sweet spot of things other than bats, you may wish to consider nonsporting equipment that also has sweet spots. Axes, for example, may have a sweet spot for most effective wood splitting. Also, scientists have determined that the extinct giant armadillo had a giant spike on its tail that was located at the sweet spot (Viegas 2009).

Conclusion

One hundred and fifty years ago, when baseball was in its infancy, no one would have dreamed that someday engineers would be employed designing sports equipment. What the future holds for the engineering of everyday items and products is also unknown. Students may have stereotypical misconceptions about the field of engineering that can be addressed by engaging them in everyday engineering pursuits.

References

Doherty, P. 2007. The node of vibration. *http://isaac. exploratorium.edu/~pauld/activities/baseball/ batnode.htm.*

International Technology Education Association (ITEA). 2002. *Standards for technological literacy: Content for the study of technology.* 2nd ed. Reston, VA: ITEA.

National Research Council (NRC). 1996. *National science education standards.* Washington, DC: National Academies Press.

University of Colorado, College of Engineering. 2006. Lesson: Engineering in sports. *www.teachengineer ing.org/view_lesson.php?url=collection/cub_/lessons/ cub_intro/cub_intro_lesson04.xml.*

Viegas, J. 2009. Prehistoric mammal swung tail like a baseball bat. *www.msnbc.msn.com/id/32559090/ns/ technology_and_science-science.*

Resources

Moyer, R., J. Hackett, and S. Everett. 2007. *Teaching science as investigations: Modeling inquiry through learning cycle lessons.* Upper Saddle River, NJ; Pearson/Merrill/Prentice Hall.

Russell, D.A. 2009. Physics and acoustics of baseball and softball bats. *http://paws.kettering.edu/~drussell/ bats.html.*

Testing bats. *www.thefutureschannel.com/dockets/ realworld/testing_bats/index.php.*

ACTIVITY WORKSHEET 14.1 — Investigating Sweet Spots

In this activity, you will investigate where on a bat you should hit a ball.

Engage

1. Think about your experiences playing baseball or softball. Have you ever hit a ball and had the bat sting your hands? Record your experiences and share with your classmates.

2. Predict what you will see, feel, and hear as one person taps a bat with a small hammer while another holds the bat by the handle end. Explain your predictions.

3. What were your observations? How do they compare to your predictions?

Explore

1. Does where you hit a ball on a bat affect how far the ball will go?

2. To answer the above question, you will use a meterstick for a bat and two taped-together nickels to represent a ball. You will hit the nickels at various points along the meterstick and measure how far the nickels travel along the tabletop.

3. Design a procedure you will follow and have your teacher approve your plan before you begin. Be sure your plan includes multiple trials.

4. Record your data in a table and calculate the average distance for each location you tested.

Explain

1. What are your conclusions?

2. How do your results compare with your classmates'?

Extend

1. How does tapping a meterstick with a hammer compare to tapping the bat?

2. Were you able to locate a sweet spot on the meterstick? If so, where?

3. How does the location of the meterstick sweet spot compare with your results in the Explore stage?

Evaluate

Your teacher will give you a hammer and a golf club, tennis racket, or other piece of sporting equipment. Explain how you will determine the location of the sweet spot for your equipment. What observations will you need to make? Have your teacher check your plan and then try it out. Explain your findings in writing.

CHAPTER 15

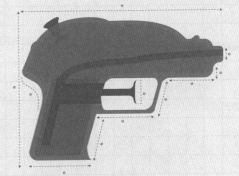

WHAT MAKES A SQUIRT GUN SQUIRT?

YOU MAY NOT expect to find engineering and squirt guns in the same sentence. However, like many examples of engineering design, the squirt gun pump mechanism is uncomplicated, yet elegant, and very inexpensive to manufacture. The squirt guns shown in Figures 15.1–15.3 were purchased at a dollar store for 33 cents apiece. The type of pump used in squirt guns is known as a positive displacement pump. Positive displacement pumps are so called because fluid is trapped within the pump and then moved through—or displaced—in one (positive) direction. The design is widely used because of its simplicity and low cost. With only a few moving parts, it is able to deliver a stream of water, a spray of cleanser, or a squirt of liquid soap.

One of our students once substituted an empty window spray bottle when his squirt gun broke. It worked well and had the added advantage of a relatively large water reservoir. Actually, the pumping mechanism of spray bottles, liquid soap dispensers, and squirt guns is essentially the same. As noted in the ITEEA standards, "A product, system, or environment developed for one setting may be applied to another setting" (ITEA 2002, p. 49). In this lesson, we will examine how these simple, everyday pumps operate.

Historical Information

The first squirt guns were developed as toys in the late 1890s. They made use of a metal toy gun with a long tube that was attached to a squeeze bulb filled with water. To operate the gun, one merely squeezed the bulb. Trigger-type squirt guns were developed in the 1930s and were the main type of water gun until the 1980s, when Super Soaker types were introduced.

The same pump technology was used for a number of other purposes. While liquid soap had been around for some time, it was not until the 1940s that the first mechanical dispensers were produced (Kleinman 2003). Aerosol dispensers require a compressed propellant and therefore must be packaged in cylindrical containers, while pump dispensers can be made in any shaped package. The propellants (chlorofluorocarbons, or CFCs) used in aerosol cans in the past were harmful to the Earth's ozone layer. In 1979, a liquid soap known as Softsoap was introduced and immediately became popular. Since 2003, foaming liquid soaps have become the latest fad. They make use of the same basic pump, but add air to the soap, which produces the foam.

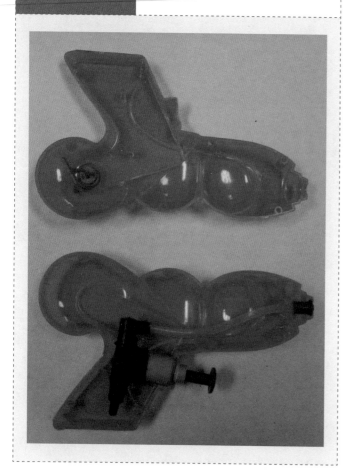

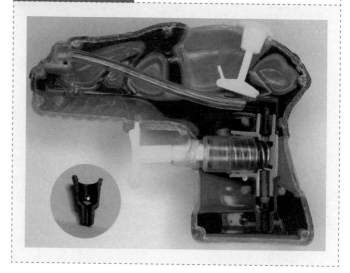

Investigating a Squirt Gun: What Makes It Squirt? (Teacher Background Information)

Engage

Safety note: Students should wear chemical-splash goggles for this entire activity.

Distribute one eyedropper and a cup of water to each group of three or four students. Only a small amount of water should be used: 3 oz. (90 ml) disposable cups partially filled. It is also recommended that student tables be covered with a bath towel. Ask students to see if they can determine how water is drawn into and pushed out of the dropper. Have stu-

dents explain in their journals what they had to do to operate the dropper (they must squeeze the bulb and then release the bulb under the surface of the water). Use this discussion to lead to the following Explore question: What makes a squirt gun work?

Explore

You will need one squirt gun for each group of three to four students. Prior to class time, you should remove the pump assembly from each squirt gun. This can be done by carefully prying open the two halves of the body of the squirt gun with a slender screwdriver (Figure 15.1). You may need to cut through the glue holding the molded sides together. Once opened, the pumping assembly can be removed intact. The squirt guns should readily come apart. Keep the parts from each squirt gun in a clear zipper-type baggie. If none of the parts are lost or broken, the pumping mechanism can be reassembled and used over again with another class of students. You may wish to have a few extra squirt guns available in case some of the small parts are lost.

FIGURE 15.3	Pumping mechanism (trigger in)

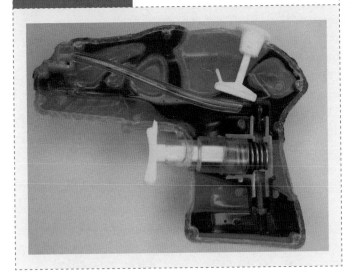

FIGURE 15.4	Liquid soap dispenser pump

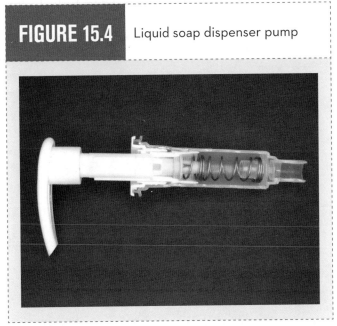

Have students determine which end must be placed in the cup of water in order for it to squirt. One end will draw water in and the other squirts it out of the pump. If students put the squirting end in the water, the pump will not work. Focus students on trying to answer the question, "What makes the gun squirt and how does that compare with how the dropper works?" As can be seen in the squirt gun in Figure 15.2, the pumping mechanism is actually made of just a few parts. The trigger pushes in a piston and compresses a spring. The body of the pump has openings at each end. There are two valves, one at each end of the pump body, and they are often called check valves. A check valve is simply a one-way valve that allows fluids to move through in only one direction. In our diagrams they are both mushroom-shaped stem valves (some pumps may have a valve made out of a small ball at the bottom—see inset photo in Figure 15.2). Note that your valves might vary but there will be two of them. There is a tube at the top of the pump body that leads to the nozzle and a short tube at the bottom to the reservoir.

Explain

A major difference between the squirt gun pump and the rudimentary pump of the eyedropper is that the dropper takes in water and expels it through the same end. When the bulb is squeezed, some air is forced out of the dropper. Therefore, the pressure in the dropper is reduced; when the bulb is released under water, the higher atmospheric pressure forces water into the dropper. In the late 1800s, squirt guns were similar to a dropper in that there was a bulb that was squeezed for its operation. The squirt gun pump is a mechanism that moves water through itself in only one direction. It draws water in one end (when the trigger is released) and expels it through the nozzle end when the trigger is depressed. How does this work? Let's consider the process step by step. The first time the trigger is depressed, air is forced out of the pump. When the trigger is released, the spring forces the piston open and the pressure in the pump is reduced. This causes both valves to move toward the pump body, which causes the upper valve to seal against the body of the pump. The water entering the pump body pushes up the lower

valve. This water remains in the pump until the trigger is pulled again. When the trigger is depressed (Figure 15.3), pressure in the pump is increased, forcing the top valve up (opening it) and pushing the lower valve down (closing it); the water is then forced out of the nozzle. Therefore, when the trigger is pulled, the top valve is open and the bottom valve is closed (Figure 15.3), but when the trigger is released, the top valve is closed and the bottom valve opens (Figure 15.2). Releasing the trigger repeats the process, filling the pump with water again.

After students have taken apart the pumps, discuss their ideas regarding how the flow of water differs in an eyedropper and in the squirt gun pump. Ask students if they can determine the flow of water through the pump. Challenge them as to the purpose of the valves. At this time you may introduce vocabulary such as valve, piston, reservoir, and nozzle.

Students should have little difficulty determining the purpose of the piston, reservoir, and the nozzle, but this may be their first investigation of a valve. You can demonstrate a ball valve by using a tornado tube (a plastic device that connects two plastic soda bottles and allows water to move from one to the other), two plastic soft drink bottles, and a marble that is just smaller than the opening of the bottle but larger than the hole in the tornado tube. Fill one bottle with water and put the marble in the other and connect the bottles with the tornado tube. Show students that the water will flow easily from one bottle to the other. Once the bottle with the marble is full of water, tip the bottle over once more and note what happens. The water may start to flow, but the marble fills the opening shutting off the flow of water. This is essentially how a ball valve in a positive displacement pump mechanism works.

Extend

Provide each group of students with the pumping mechanism from a liquid soap dispenser or a spray

bottle. If you reuse a cleanser bottle, make sure that it has been thoroughly rinsed. Empty bottles can also be purchased at most dollar stores for approximately $1 each. Students should conclude that although they may look a bit different, these pumps function in the same way as those found in squirt guns. They all have some type of piston pump, a reservoir of liquid, a nozzle of some sort, and two valves (Figure 15.4, p. 111). The valves may differ—you may find a flap, a disk, or other shapes. Note that once a device has been engineered, it can often be used, with minor changes, for many other purposes—in this case, everything from squirt guns to soap dispensers to spray bottles. You can ask students to find examples at home and share the results of this type of scavenger hunt with the class. Another principle of engineering also shown here is that designers have been able to make many everyday devices with very few moving parts and for very low manufacturing costs.

Evaluate

Students should be able to make a sketch of the critical parts of their pumping mechanism from the Extend stage. They should label and indicate with arrows the flow of liquid. Each sketch should include a reservoir, a pump with a spring and piston, a nozzle, and two one-way valves.

Conclusion

A basic principle of engineering is to apply known technology to new applications. In this lesson, students investigate several uses for inexpensive positive displacement pumps. They also have the opportunity to try to invent their own use for such devices. This encourages students to become curious about how even the simple things around them function. This curiosity may be the first step for students to develop an interest in engineering as a possible career.

References

International Technology Education Association. 2002. *Standards for technological literacy: Content for the study of technology.* 2nd ed. Reston, VA: ITEA.

Kleinman, M. 2003. New life in the handsoap. *Soap and cosmetics, a Chemical Week Associates publication.* February.

Moyer, R., J. Hackett, and S. Everett. 2007. *Teaching science as investigations: Modeling inquiry through learning cycle lessons.* Upper Saddle River, NJ: Pearson/Merrill/Prentice Hall.

ACTIVITY WORKSHEET 15.1 Investigating a Squirt Gun: What Makes It Squirt?

In this activity, you are going to take apart a squirt gun to find out what makes it squirt and compare it with an eyedropper.

Engage

Safety note: Wear chemical-splash goggles for this activity.

1. Cover your work area with a towel or newspapers. Using the materials from your teacher, fill and empty the dropper to see if you can determine how it works.

2. What must you do to fill it with water? What must you do to empty the water?

Explore

1. Examine the pumping mechanism from the squirt gun. What must you do to fill and empty the pumping mechanism with water?

2. Carefully take apart the pumping mechanism without breaking the pieces. Try to determine how each part in this system works to draw water in and squirt it out.

3. Make a drawing of your findings to show how the squirt gun pump operates. Use arrows to show the flow of water.

Explain

1. Make a drawing of the eyedropper. Use arrows to show the flow of water in and out.

2. How does the eyedropper differ from the squirt gun pump?

3. What do you think the small parts at the top and bottom of the body of the pump are used for?

4. Your teacher has a large model of one type of valve. What do you think is the purpose of the ball?

Extend

1. Observe the pump your teacher has provided. For what was your pump used?

2. Is this pump more like the eyedropper or the squirt gun?

3. Does this pump have any valves? If so, where are they located?

4. Brainstorm other uses for the positive displacement pump. Describe what task your invention accomplishes.

Evaluate

Draw and label the pump and the flow of liquid.

INDEX

*Page numbers printed in **boldface** type refer to figures.*

INDEX